OFFICIAL SQA PAST PAPERS

WITH ANSWERS

HIGHER

CHEMISTRY
2008-2012

© Scottish Qualifications Authority

First exam published in 2008.
Published by Bright Red Publishing Ltd, 6 Stafford Street, Edinburgh EH3 7AU
tel: 0131 220 5804 fax: 0131 220 6710 info@brightredpublishing.co.uk www.brightredpublishing.co.uk

ISBN 978-1-84948-284-4

A CIP Catalogue record for this book is available from the British Library.

Bright Red Publishing is grateful to the copyright holders, as credited on the final page of the Question Section, for permission to use their material. Every effort has been made to trace the copyright holders and to obtain their permission for the use of copyright material. Bright Red Publishing will be happy to receive information allowing us to rectify any error or omission in future editions.

HIGHER

2008

[BLANK PAGE]

FOR OFFICIAL USE

Total
Section B

X012/301

NATIONAL QUALIFICATIONS 2008

FRIDAY, 30 MAY
9.00 AM – 11.30 AM

CHEMISTRY
HIGHER

Fill in these boxes and read what is printed below.

Full name of centre

Town

Forename(s)

Surname

Date of birth

Day Month Year   Scottish candidate number   Number of seat

Reference may be made to the Chemistry Higher and Advanced Higher Data Booklet.

SECTION A–Questions 1—40 (40 marks)

Instructions for completion of **Section A** are given on page two.

For this section of the examination you must use an **HB pencil**.

SECTION B (60 marks)

1   All questions should be attempted.

2   The questions may be answered in any order but all answers are to be written in the spaces provided in this answer book, **and must be written clearly and legibly in ink**.

3   Rough work, if any should be necessary, should be written in this book and then scored through when the fair copy has been written. If further space is required, a supplementary sheet for rough work may be obtained from the invigilator.

4   Additional space for answers will be found at the end of the book. If further space is required, supplementary sheets may be obtained from the invigilator and should be inserted inside the **front** cover of this book.

5   The size of the space provided for an answer should not be taken as an indication of how much to write. It is not necessary to use all the space.

6   Before leaving the examination room you must give this book to the invigilator. If you do not, you may lose all the marks for this paper.

SECTION A

Read carefully

1 Check that the answer sheet provided is for **Chemistry Higher (Section A)**.

2 For this section of the examination you must use an **HB pencil** and, where necessary, an eraser.

3 Check that the answer sheet you have been given has **your name**, **date of birth**, **SCN** (Scottish Candidate Number) and **Centre Name** printed on it.

    Do not change any of these details.

4 If any of this information is wrong, tell the Invigilator immediately.

5 If this information is correct, **print** your name and seat number in the boxes provided.

6 The answer to each question is **either** A, B, C or D. Decide what your answer is, then, using your pencil, put a horizontal line in the space provided (see sample question below).

7 There is **only one correct** answer to each question.

8 Any rough working should be done on the question paper or the rough working sheet, **not** on your answer sheet.

9 At the end of the exam, put the **answer sheet for Section A inside the front cover of your answer book**.

Sample Question

To show that the ink in a ball-pen consists of a mixture of dyes, the method of separation would be

      A chromatography

      B fractional distillation

      C fractional crystallisation

      D filtration.

The correct answer is **A**—chromatography. The answer **A** has been clearly marked in **pencil** with a horizontal line (see below).

Changing an answer

If you decide to change your answer, carefully erase your first answer and using your pencil, fill in the answer you want. The answer below has been changed to **D**.

      A   B   C   D

1. Solutions of barium chloride and silver nitrate are mixed together.

   The reaction that takes place is an example of

   A displacement

   B neutralisation

   C oxidation

   D precipitation.

2. Two rods are placed in dilute sulphuric acid as shown.

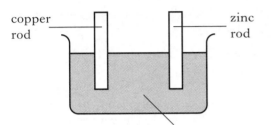

   copper rod   zinc rod

   dilute sulphuric acid

   Which of the following would be observed?

   A No gas is given off.

   B Gas is given off at only the zinc rod.

   C Gas is given off at only the copper rod.

   D Gas is given off at both rods.

3. An element was burned in air. The product was added to water, producing a solution with a pH less than 7.

   The element could be

   A carbon

   B hydrogen

   C sodium

   D tin.

4. A mixture of sodium chloride and sodium sulphate is known to contain 0·6 mol of chloride ions and 0·2 mol of sulphate ions.

   How many moles of sodium ions are present?

   A 0·4

   B 0·5

   C 0·8

   D 1·0

5. The following results were obtained in the reaction between marble chips and dilute hydrochloric acid.

Time/minutes	0	2	4	6	8	10
Total volume of carbon dioxide produced/cm^3	0	52	68	78	82	84

   What is the average rate of production of carbon dioxide, in $cm^3\ min^{-1}$, between 2 and 8 minutes?

   A 5

   B 26

   C 30

   D 41

6. 5 g of copper is added to excess silver nitrate solution. The equation for the reaction that takes place is:

   $$Cu(s) + 2AgNO_3(aq) \rightarrow 2Ag(s) + Cu(NO_3)_2(aq)$$

   After some time, the solid present is filtered off from the solution, washed with water, dried and weighed.

   The final mass of the solid will be

   A less than 5 g

   B 5 g

   C 10 g

   D more than 10 g.

7.

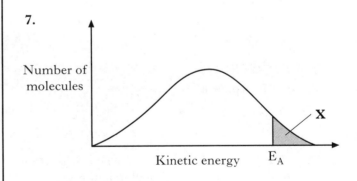

   Number of molecules

   Kinetic energy E_A

   In area **X**

   A molecules always form an activated complex

   B no molecules have the energy to form an activated complex

   C collisions between molecules are always successful in forming products

   D all molecules have the energy to form an activated complex.

8.

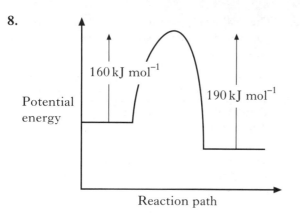

Reaction path

When a catalyst is used, the activation energy of the forward reaction is reduced to $35\,kJ\,mol^{-1}$.

What is the activation energy of the catalysed reverse reaction?

A $30\,kJ\,mol^{-1}$

B $35\,kJ\,mol^{-1}$

C $65\,kJ\,mol^{-1}$

D $190\,kJ\,mol^{-1}$

9. As the atomic number of the alkali metals increases

A   the first ionisation energy decreases

B   the atomic size decreases

C   the density decreases

D   the melting point increases.

10. Which of the following atoms has the least attraction for bonding electrons?

A   Carbon

B   Nitrogen

C   Phosphorus

D   Silicon

11. Which of the following reactions refers to the third ionisation energy of aluminium?

A $Al(s) \rightarrow Al^{3+}(g) + 3e^-$

B $Al(g) \rightarrow Al^{3+}(g) + 3e^-$

C $Al^{2+}(g) \rightarrow Al^{3+}(g) + e^-$

D $Al^{3+}(g) \rightarrow Al^{4+}(g) + e^-$

12. Which of the following represents an exothermic process?

A $Cl_2(g) \rightarrow 2Cl(g)$

B $Na(s) \rightarrow Na(g)$

C $Na(g) \rightarrow Na^+(g) + e^-$

D $Na^+(g) + Cl^-(g) \rightarrow Na^+Cl^-(s)$

13. In which of the following liquids does hydrogen bonding occur?

A   Ethanol

B   Ethyl ethanoate

C   Hexane

D   Pent-1-ene

14. The shapes of some common molecules are shown. Each molecule contains at least one polar covalent bond.

Which of the following molecules is non-polar?

A $H — Cl$

B
$$O$$
$$H \quad \quad H$$

C $O = C = O$

D
$$H$$
$$Cl \quad C \quad Cl$$
$$Cl$$

15. At room temperature, a solid substance was shown to have a lattice consisting of positively charged ions and delocalised outer electrons.

The substance could be

A   graphite

B   sodium

C   mercury

D   phosphorus.

16. The mass of 1 mol of sodium is 23 g.

What is the approximate mass of one sodium atom?

A 6×10^{23} g

B 6×10^{-23} g

C $3 \cdot 8 \times 10^{-23}$ g

D $3 \cdot 8 \times 10^{-24}$ g

17. In which of the following pairs do the gases contain the same number of oxygen atoms?

A 1 mol of oxygen and 1 mol of carbon monoxide

B 1 mol of oxygen and 0·5 mol of carbon dioxide

C 0·5 mol of oxygen and 1 mol of carbon dioxide

D 1 mol of oxygen and 1 mol of carbon dioxide

18. The Avogadro Constant is the same as the number of

A molecules in 16 g of oxygen

B electrons in 1 g of hydrogen

C atoms in 24 g of carbon

D ions in 1 litre of sodium chloride solution, concentration $1 \, mol \, l^{-1}$.

19. $$2NO(g) + O_2(g) \rightarrow 2NO_2(g)$$

How many litres of nitrogen dioxide gas would be produced in a reaction, starting with a mixture of 5 litres of nitrogen monoxide gas and 2 litres of oxygen gas?

(All volumes are measured under the same conditions of temperature and pressure.)

A 2

B 3

C 4

D 5

20. Which of the following fuels can be produced by the fermentation of biological material under anaerobic conditions?

A Hydrogen

B Methane

C Methanol

D Petrol

21. Butadiene is the first member of a homologous series of hydrocarbons called dienes.

What is the general formula for this series?

A C_nH_{n+2}

B C_nH_{n+3}

C C_nH_{2n}

D C_nH_{2n-2}

22.

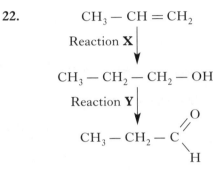

Which line in the table correctly describes reactions **X** and **Y**?

	Reaction X	Reaction Y
A	hydration	oxidation
B	hydration	reduction
C	hydrolysis	oxidation
D	hydrolysis	reduction

23. Ammonia is manufactured from hydrogen and nitrogen by the Haber Process.

$$3H_2(g) + N_2(g) \rightleftharpoons 2NH_3(g)$$

If 80 kg of ammonia is produced from 60 kg of hydrogen, what is the percentage yield?

A $\dfrac{80}{340} \times 100$

B $\dfrac{80}{170} \times 100$

C $\dfrac{30}{80} \times 100$

D $\dfrac{60}{80} \times 100$

[Turn over

24. Which of the following statements about methanol is **false**?

    A   It can be made from synthesis gas.

    B   It can be dehydrated to form an alkene.

    C   It can be oxidised to give a carboxylic acid.

    D   It reacts with acidified potassium dichromate solution.

25. A by-product produced in the manufacture of a polyester has the structure shown.

    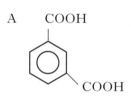

    What is the structure of the diacid monomer used in the polymerisation?

    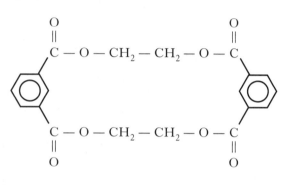

    A   COOH

    B   COOH

    C   HOOC — CH₂ — CH₂ — COOH

    D   COOH

26. Which of the following statements can be applied to polymeric esters?

    A   They are used for flavourings, perfumes and solvents.

    B   They are manufactured for use as textile fibres and resins.

    C   They are cross-linked addition polymers.

    D   They are condensation polymers made by the linking up of amino acids.

27. The rate of hydrolysis of protein, using an enzyme, was studied at different temperatures.

    Which of the following graphs would be obtained?

    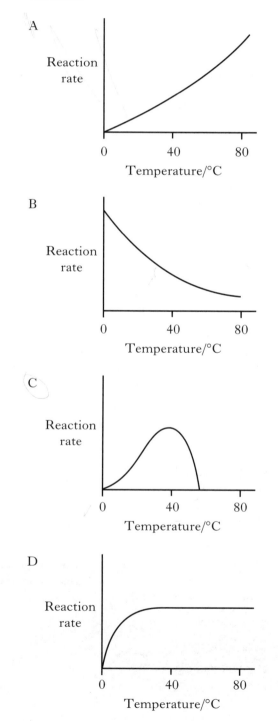

28. Which of the following arrangements of atoms shows a peptide link?

$$
\begin{array}{c}
\qquad \overset{\displaystyle H}{\underset{\displaystyle H}{|}} \qquad \overset{\displaystyle H}{|} \\
A \quad -C-O-N- \\
\qquad |
\end{array}
$$

A
```
     H        H
     |        |
  — C — O — N —
     |
     H
```

B
```
     H    O    H
     |    ||   |
  — C  — C — N —
     |
     H
```

C
```
     H    OH
     |    |
  — C  — C = N —
     |
     H
```

D
```
     H    O        H
     |    ||       |
  — C  — C — O — N —
     |
     H
```

29. Which line in the table shows the effect of a catalyst on the reaction rates and position of equilibrium in a reversible reaction?

	Rate of forward reaction	Rate of reverse reaction	Position of equilibrium
A	increased	unchanged	moves right
B	increased	increased	unchanged
C	increased	decreased	moves right
D	unchanged	unchanged	unchanged

30. The following equilibrium exists in bromine water.

$$Br_2(aq) + H_2O(\ell) \rightleftharpoons Br^-(aq) + 2H^+(aq) + OBr^-(aq)$$
(red)   (colourless)   (colourless)

The red colour of bromine water would fade on adding a few drops of a concentrated solution of

A HCl

B KBr

C $AgNO_3$

D NaOBr.

31. Which of the following is the best description of a $0 \cdot 1 \ mol \, l^{-1}$ solution of nitric acid?

A Dilute solution of a weak acid

B Dilute solution of a strong acid

C Concentrated solution of a weak acid

D Concentrated solution of a strong acid

32. The conductivity of pure water is low because

A water molecules are polar

B only a few water molecules are ionised

C water molecules are linked by hydrogen bonds

D there are equal numbers of hydrogen and hydroxide ions in water.

33. Which of the following statements is **true** about an aqueous solution of ammonia?

A It has a pH less than 7.

B It is completely ionised.

C It contains more hydroxide ions than hydrogen ions.

D It reacts with acids producing ammonia gas.

34. Equal volumes of solutions of ethanoic acid and hydrochloric acid, of equal concentration, are compared.

In which of the following cases does the ethanoic acid give the higher value?

A pH of solution

B Conductivity of solution

C Rate of reaction with magnesium

D Volume of sodium hydroxide solution neutralised

35. Equal volumes of $0 \cdot 1 \ mol \, l^{-1}$ solutions of the following acids and alkalis were mixed.

Which of the following pairs would give the solution with the lowest pH?

A Hydrochloric acid and sodium hydroxide

B Hydrochloric acid and calcium hydroxide

C Sulphuric acid and sodium hydroxide

D Sulphuric acid and calcium hydroxide

[Turn over

36. Which of the following compounds dissolves in water to form an acidic solution?

   A   Sodium nitrate

   B   Barium sulphate

   C   Potassium ethanoate

   D   Ammonium chloride

37. The ion-electron equations for a redox reaction are:

$$2I^-(aq) \rightarrow I_2(aq) + 2e^-$$

$$MnO_4^-(aq) + 8H^+(aq) + 5e^- \rightarrow Mn^{2+}(aq) + 4H_2O(\ell)$$

   How many moles of iodide ions are oxidised by one mole of permanganate ions?

   A   0·2

   B   0·4

   C   2

   D   5

38. In which of the following reactions is the hydrogen ion acting as an oxidising agent?

   A $Mg + 2HCl \rightarrow MgCl_2 + H_2$

   B $NaOH + HNO_3 \rightarrow NaNO_3 + H_2O$

   C $CuCO_3 + H_2SO_4 \rightarrow CuSO_4 + H_2O + CO_2$

   D $CH_3COONa + HCl \rightarrow NaCl + CH_3COOH$

39. An atom of ^{227}Th decays by a series of alpha emissions to form an atom of ^{211}Pb.

   How many alpha particles are released in the process?

   A   2

   B   3

   C   4

   D   5

40. The half-life of the isotope ^{210}Pb is 21 years.

   What fraction of the original ^{210}Pb atoms will be present after 63 years?

   A   0·5

   B   0·25

   C   0·125

   D   0·0625

Candidates are reminded that the answer sheet MUST be returned INSIDE the front cover of this answer book.

Marks

SECTION B

All answers must be written clearly and legibly in ink.

1. The formulae for three oxides of sodium, carbon and silicon are Na_2O, CO_2 and SiO_2.

   Complete the table for CO_2 and SiO_2 to show both the bonding and structure of the three oxides at room temperature.

Oxide	Bonding and structure
Na_2O	ionic lattice
CO_2	covelent molecular
SiO_2	covdlent network

   (2)

2. A typical triglyceride found in olive oil is shown below.

   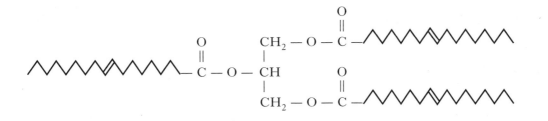

   (a) To which family of organic compounds do triglycerides belong?

   esters

   1

   (b) Olive oil can be hardened for use in margarines.
   What happens to the triglyceride molecules during the hardening of olive oil?

   1

   (c) Give **one** reason why oils can be a useful part of a balanced diet.

   loners chlroesdrol

   heanny

   1
   (3)

Marks

3. A student carried out the Prescribed Practical Activity (PPA) to find the effect of concentration on the rate of the reaction between hydrogen peroxide solution and an acidified solution of iodide ions.

$$H_2O_2(aq) \quad + \quad 2H^+(aq) \quad + \quad 2I^-(aq) \quad \rightarrow \quad 2H_2O(\ell) \quad + \quad I_2(aq)$$

During the investigation, only the concentration of the iodide ions was changed.

Part of the student's results sheet for this PPA is shown.

Results

Experiment	Volume of KI(aq) /cm^3	Volume of H$_2$O /cm^3	Volume of H$_2$O$_2$(aq) /cm^3	Volume of H$_2$SO$_4$(aq) /cm^3	Volume of Na$_2$S$_2$O$_3$(aq) /cm^3	Rate /s^{-1}
1	25	0	5	10	10	0·043
2						
3						

(a) Describe how the concentration of the potassium iodide solution was changed during this series of experiments.

1

(b) Calculate the reaction time, in seconds, for the first experiment.

$R = \dfrac{1}{t}$

1

(2)

Marks

4. Using a cobalt catalyst, alkenes react with a mixture of hydrogen and carbon monoxide.

The products are two isomeric aldehydes.

Propene reacts with the mixture as shown.

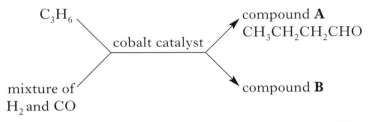

C_3H_6

cobalt catalyst

mixture of
H_2 and CO

compound **A**
$CH_3CH_2CH_2CHO$

compound **B**

(a) What name is given to a mixture of hydrogen and carbon monoxide?

1

(b) Draw a structural formula for compound **B**.

1

(c) (i) What would be observed if compound **A** was gently heated with Tollens' reagent?

1

(ii) How would the reaction mixture be heated?

1

(d) Aldehydes can also be formed by the reaction of some alcohols with copper(II) oxide.

Name the **type** of alcohol that would react with copper(II) oxide to form an aldehyde.

primary alcohol

1

(5)

[Turn over

Marks

5. All the isotopes of technetium are radioactive.

(*a*) Technetium-99 is produced as shown.

$$^{99}_{42}\text{Mo} \rightarrow ^{99}_{43}\text{Tc} + \mathbf{X}$$

Identify **X**.

1

(*b*) The graph shows the decay curve for a 1·0 g sample of technetium-99.

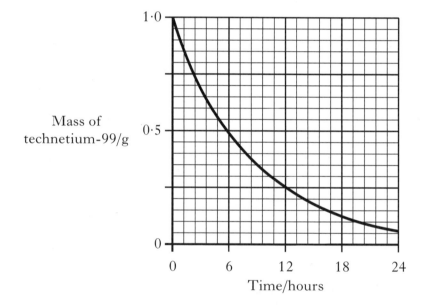

Mass of
technetium-99/g

Time/hours

(i) Draw a curve on the graph to show the variation of mass with time for a 0·5 g sample of technetium-99.

(An additional graph, if required, can be found on *Page twenty-eight*.)

1

(ii) Technetium-99 is widely used in medicine to detect damage to heart tissue. It is a gamma-emitting radioisotope and is injected into the body.

Suggest **one** reason why technetium-99 can be safely used in this way.

1

(3)

Marks

6. In 1865, the German chemist Kekulé proposed a ring structure for benzene. This structure was based on alternating single and double bonds.

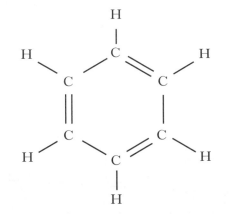

(a) (i) Describe a chemical test that would indicate that the above chemical structure for benzene is incorrect.

decolours bromine solution doesn't rapidly

1

(ii) Briefly describe the correct structure for benzene.

delocorsed e

1

(b) Benzene can be formed from cyclohexane.

$$C_6H_{12} \rightarrow C_6H_6 + 3H_2$$

What name is given to this type of reaction?

reforming

1

(c) Benzene is also added in very small amounts to some petrols.
Why is benzene added to petrol?

increases octane number

1

(4)

Marks

7. Hydrogen fluoride, HF, is used to manufacture hydrofluorocarbons.

Hydrofluorocarbons are now used as refrigerants instead of chlorofluorocarbons, CFCs.

(a) Why are CFCs no longer used?

depletion of ozone

1

(b) Hydrogen fluoride gas is manufactured by reacting calcium fluoride with concentrated sulphuric acid.

$$CaF_2 + H_2SO_4 \rightarrow CaSO_4 + 2HF$$

What volume of hydrogen fluoride gas is produced when $1\cdot0$ kg of calcium fluoride reacts completely with concentrated sulphuric acid?

(Take the molar volume of hydrogen fluoride gas to be 24 litres mol^{-1}.)

Show your working clearly.

1mol — 2mol.

78g — 48   24×2
                =48

1000 — x

78x = 48000

x = 48000 / 78

= 615.38

CaF_2
40
19×2
=38g

2HF

19
=21

2
(3)

Marks

8. Carbon monoxide can be produced in many ways.

(a) One method involves the reaction of carbon with an oxide of boron.

$$2B_2O_3 \quad + \quad 7C \quad \rightarrow \quad B_4C \quad + \quad 6CO$$

Balance this equation.

1

(b) Carbon monoxide is also a product of the reaction of carbon dioxide with hot carbon. The carbon dioxide is made by the reaction of dilute hydrochloric acid with solid calcium carbonate.

Unreacted carbon dioxide is removed before the carbon monoxide is collected by displacement of water.

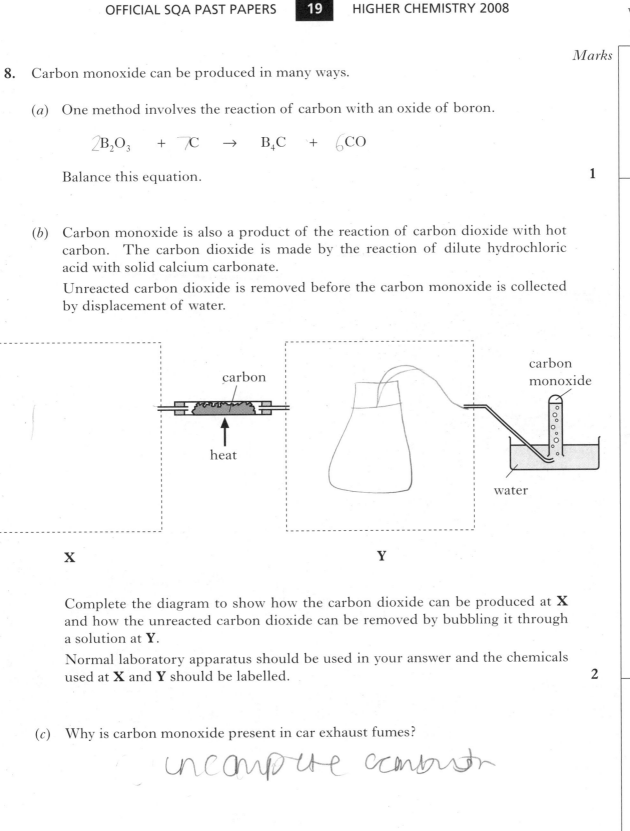

X   **Y**

Complete the diagram to show how the carbon dioxide can be produced at **X** and how the unreacted carbon dioxide can be removed by bubbling it through a solution at **Y**.

Normal laboratory apparatus should be used in your answer and the chemicals used at **X** and **Y** should be labelled.

2

(c) Why is carbon monoxide present in car exhaust fumes?

incomplete combustion

1

(4)

[Turn over

Marks

9. Hydrogen gas has a boiling point of −253 °C.

(*a*) Explain clearly why hydrogen is a gas at room temperature.

In your answer you should name the intermolecular forces involved and indicate how they arise.

2

(*b*) Hydrogen gas can be prepared in the lab by the electrolysis of dilute sulphuric acid.

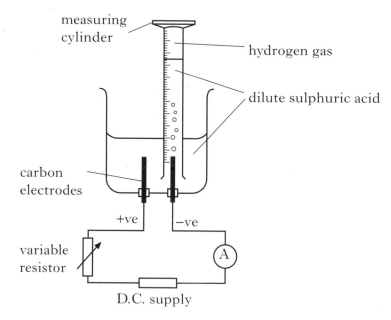

(i) Before collecting the gas in the measuring cylinder, it is usual to switch on the current and allow bubbles of gas to be produced for a few minutes.

Why is this done?

1

Marks

9. **(b)** **(continued)**

(ii) The equation for the reaction at the negative electrode is:

$$2H^+(aq) + 2e^- \rightarrow H_2(g)$$

Calculate the mass of hydrogen gas, in grams, produced in 10 minutes when a current of $0.30\,A$ was used.

Show your working clearly.

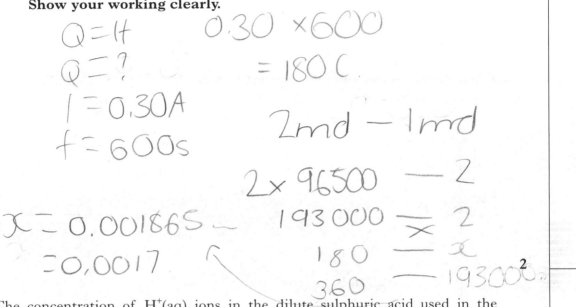

$Q = It$
$Q = ?$
$I = 0.30\,A$
$t = 600\,s$

0.30×600
$= 180\,C$

$2\,mol - 1\,mol$

$2 \times 96500 \quad - 2$
$193\,000 = 2$
$180 \quad - x$
$360 \quad - 193\,000$

$x = 0.001865$
$= 0.0017$

2

(c) The concentration of $H^+(aq)$ ions in the dilute sulphuric acid used in the experiment was $1 \times 10^{-1}\,mol\,l^{-1}$.

Calculate the concentration of $OH^-(aq)$ ions, in $mol\,l^{-1}$, in the dilute sulphuric acid.

1×10^{-13}

1

(6)

[Turn over

Marks

10. When cyclopropane gas is heated over a catalyst, it isomerises to form propene gas and an equilibrium is obtained.

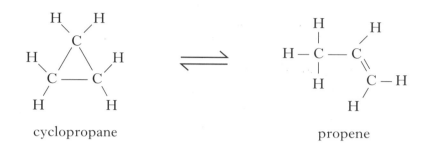

          cyclopropane   propene

The graph shows the concentrations of cyclopropane and propene as equilibrium is established in the reaction.

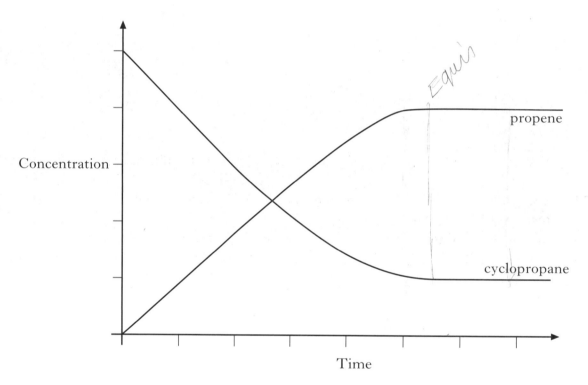

(a) Mark clearly on the graph the point at which equilibrium has just been reached.

1

(b) Why does increasing the pressure have **no** effect on the position of this equilibrium?

1

Marks

10.   (continued)

   (*c*)   The equilibrium can also be achieved by starting with propene.

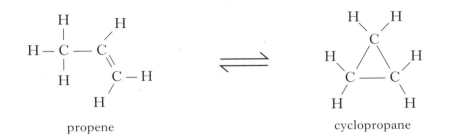

propene   cyclopropane

Using the initial concentrations shown, sketch a graph to show how the concentrations of propene and cyclopropane change as equilibrium is reached for this reverse reaction.

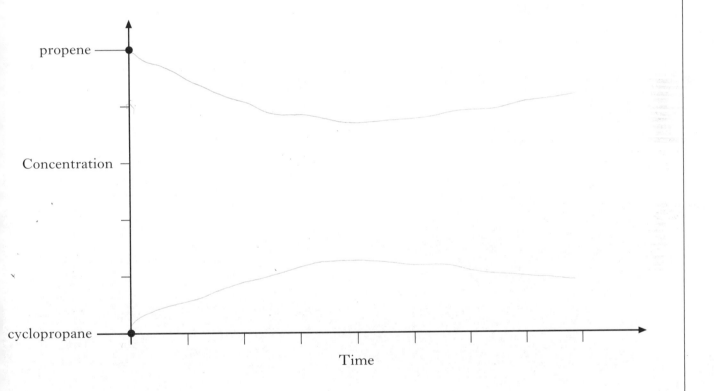

1
(3)

[Turn over

Marks

11. A student writes the following two statements. **Both are incorrect**.
    In each case explain the mistake in the student's reasoning.

    (a) Alcohols are alkaline because of their OH groups.

    1

    (b) Because of the iodine, potassium iodide will produce a blue/black colour in contact with starch.

    1
    (2)

Marks

12. When in danger, bombardier beetles can fire a hot, toxic mixture of chemicals at the attacker.

    This mixture contains quinone, $C_6H_4O_2$, a compound that is formed by the reaction of hydroquinone, $C_6H_4(OH)_2$, with hydrogen peroxide, H_2O_2. The reaction is catalysed by an enzyme called catalase.

    (a) Most enzymes can catalyse only specific reactions, eg catalase cannot catalyse the hydrolysis of starch.

    Give a reason for this.

    The shape of the substrate
    will only bind on type of
    enzyme

    1

    (b) The equation for the overall reaction is:

    $$C_6H_4(OH)_2(aq) \quad + \quad H_2O_2(aq) \quad \rightarrow \quad C_6H_4O_2(aq) \quad + \quad 2H_2O(\ell)$$

    Use the following data to calculate the enthalpy change, in kJ mol⁻¹, for the above reaction.

    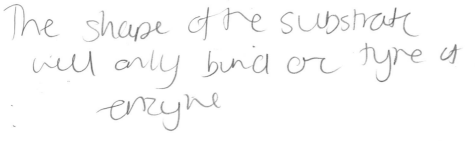

    $$C_6H_4(OH)_2(aq) \quad \rightarrow \quad C_6H_4O_2(aq) \quad + \quad H_2(g) \qquad \Delta H = +177 \cdot 4 \, \text{kJ mol}^{-1}$$

    $$H_2(g) + O_2(g) \quad \rightarrow \quad H_2O_2(aq) \qquad \Delta H = -191 \cdot 2 \, \text{kJ mol}^{-1}$$

    $$H_2(g) + \tfrac{1}{2}O_2(g) \quad \rightarrow \quad H_2O(g) \qquad \Delta H = -241 \cdot 8 \, \text{kJ mol}^{-1}$$

    $$H_2O(g) \quad \rightarrow \quad H_2O(\ell) \qquad \Delta H = -43 \cdot 8 \, \text{kJ mol}^{-1}$$

    Show your working clearly.

    ① × 1 ② × -1 ③ × -2 ① × -1

    2

    (3)

    [Turn over

Marks

13. For many years, carbohydrates found in plants have been used to provide chemicals. Lactic acid can be produced by fermenting the carbohydrates in corn.

Lactic acid has the structure:

(a) Name the functional group in the shaded area.

hydroxyl

1

(b) Lactic acid is used to make polylactic acid, a biodegradeable polymer that is widely used for food packaging.

(i) Name another biodegradeable polymer.

biopol

1

(ii) Polylactic acid can be manufactured by either a batch or a continuous process.

What is meant by a batch process?

When reactants are feed into a reacting vessel left reacted then removed. The vessel is then cleaned

1

(iii) The first stage in the polymerisation of lactic acid involves the condensation of two lactic acid molecules to form a cyclic structure called a lactone.

Draw a structural formula for the lactone formed when two molecules of lactic acid undergo condensation with each other.

1

(4)

Marks

14. Hydrogen peroxide decomposes as shown:

$$H_2O_2(aq) \rightarrow H_2O(\ell) + \tfrac{1}{2}O_2(g)$$

The reaction can be catalysed by iron(III) nitrate solution.

(*a*) What **type** of catalyst is iron(III) nitrate solution in this reaction?

1

(*b*) In order to calculate the enthalpy change for the decomposition of hydrogen peroxide, a student added iron(III) nitrate solution to hydrogen peroxide solution.

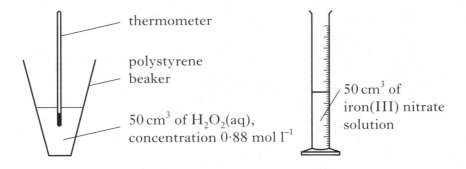

As a result of the reaction, the temperature of the solution in the polystyrene beaker increased by 16 °C.

(i) What is the effect of the catalyst on the enthalpy change (ΔH) for the reaction?

no effect

1

(ii) Use the experimental data to calculate the enthalpy change, in kJ mol^{-1}, for the decomposition of hydrogen peroxide.

Show your working clearly.

3

(5)

Marks

15. (*a*) The graph shows how the freezing point changes with changing concentration for aqueous solutions of sodium chloride and ethane-1,2-diol.

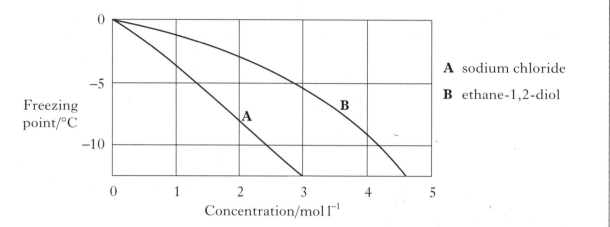

A sodium chloride

B ethane-1,2-diol

(i) Draw a structural formula for ethane-1,2-diol.

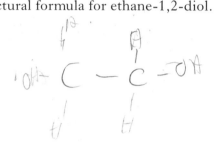

1

(ii) Ethane-1,2-diol solution is used as an antifreeze in car radiators, yet from the graph it would appear that sodium chloride solution is more efficient.

Suggest why sodium chloride solution is **not** used as an antifreeze.

cause rusting

1

(*b*) Boiling points can be used to compare the strengths of the intermolecular forces in alkanes with the strengths of the intermolecular forces in diols.

Name the alkane that should be used to make a valid comparison between the strength of its intermolecular forces and those in ethane-1,2-diol.

butane

1

(3)

Marks

16. Aldehydes and ketones can take part in a reaction sometimes known as an aldol condensation.

The simplest aldol reaction involves two molecules of ethanal.

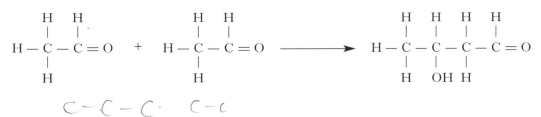

In the reaction, the carbon atom next to the carbonyl functional group of one molecule forms a bond with the carbonyl carbon atom of the second molecule.

(a) Draw a structural formula for the product formed when propanone is used instead of ethanal in this type of reaction.

1

(b) Name an aldehyde that would **not** take part in an aldol condensation.

1

(c) Apart from the structure of the reactants, suggest what is unusual about applying the term "condensation" to this particular type of reaction.

1

(3)

[Turn over

Marks

17. Oxalic acid is found in rhubarb. The number of moles of oxalic acid in a carton of rhubarb juice can be found by titrating samples of the juice with a solution of potassium permanganate, a powerful oxidising agent.

The equation for the overall reaction is:

$$5(COOH)_2(aq) + 6H^+(aq) + 2MnO_4^-(aq) \rightarrow 2Mn^{2+}(aq) + 10CO_2(aq) + 8H_2O(\ell)$$

   (a) Write the ion-electron equation for the reduction reaction.

1

   (b) Why is an indicator **not** required to detect the end-point of the titration?

1

   (c) In an investigation using a 500 cm³ carton of rhubarb juice, separate 25·0 cm³ samples were measured out. Three samples were then titrated with 0·040 mol l⁻¹ potassium permanganate solution, giving the following results.

Titration	Volume of potassium permanganate solution used/cm³
1	27·7
2	26·8
3	27·0

Average volume of potassium permanganate solution used = 26·9 cm³.

   (i) Why was the first titration result not included in calculating the average volume of potassium permanganate solution used?

1

DO NOT
WRITE IN
THIS
MARGIN

Marks

17. (*c*) **(continued)**

(ii) Calculate the number of moles of oxalic acid in the $500 \, cm^3$ carton of rhubarb juice.

Show your working clearly.

5md —

2

(5)

[*END OF QUESTION PAPER*]

SPACE FOR ANSWERS

ADDITIONAL GRAPH FOR QUESTION 5(b)(i)

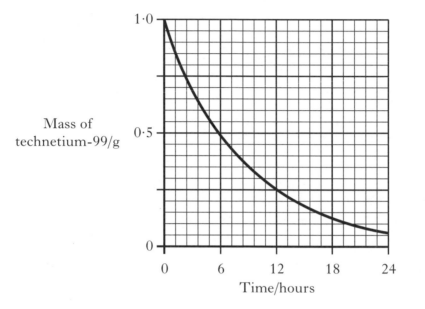

ADDITIONAL SPACE FOR ANSWERS

ADDITIONAL SPACE FOR ANSWERS

HIGHER

2009

[BLANK PAGE]

FOR OFFICIAL USE

Total
Section B

X012/301

NATIONAL
QUALIFICATIONS
2009

WEDNESDAY, 3 JUNE
9.00 AM – 11.30 AM

CHEMISTRY
HIGHER

Fill in these boxes and read what is printed below.

Full name of centre

Town

Forename(s)

Surname

Date of birth
   Day   Month   Year   Scottish candidate number   Number of seat

Reference may be made to the Chemistry Higher and Advanced Higher Data Booklet.

SECTION A–Questions 1—40 (40 marks)

Instructions for completion of **Section A** are given on page two.

For this section of the examination you must use an **HB pencil**.

SECTION B (60 marks)

1   All questions should be attempted.

2   The questions may be answered in any order but all answers are to be written in the spaces provided in this answer book, **and must be written clearly and legibly in ink**.

3   Rough work, if any should be necessary, should be written in this book and then scored through when the fair copy has been written. If further space is required, a supplementary sheet for rough work may be obtained from the invigilator.

4   Additional space for answers will be found at the end of the book. If further space is required, supplementary sheets may be obtained from the invigilator and should be inserted inside the **front** cover of this book.

5   The size of the space provided for an answer should not be taken as an indication of how much to write. It is not necessary to use all the space.

6   Before leaving the examination room you must give this book to the invigilator. If you do not, you may lose all the marks for this paper.

SECTION A

Read carefully

1 Check that the answer sheet provided is for **Chemistry Higher (Section A)**.

2 For this section of the examination you must use an **HB pencil** and, where necessary, an eraser.

3 Check that the answer sheet you have been given has **your name**, **date of birth**, **SCN** (Scottish Candidate Number) and **Centre Name** printed on it.

   Do not change any of these details.

4 If any of this information is wrong, tell the Invigilator immediately.

5 If this information is correct, **print** your name and seat number in the boxes provided.

6 The answer to each question is **either** A, B, C or D. Decide what your answer is, then, using your pencil, put a horizontal line in the space provided (see sample question below).

7 There is **only one correct** answer to each question.

8 Any rough working should be done on the question paper or the rough working sheet, **not** on your answer sheet.

9 At the end of the exam, put the **answer sheet for Section A inside the front cover of your answer book**.

Sample Question

To show that the ink in a ball-pen consists of a mixture of dyes, the method of separation would be

   A chromatography

   B fractional distillation

   C fractional crystallisation

   D filtration.

The correct answer is **A**—chromatography. The answer **A** has been clearly marked in **pencil** with a horizontal line (see below).

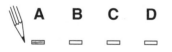

Changing an answer

If you decide to change your answer, carefully erase your first answer and using your pencil, fill in the answer you want. The answer below has been changed to **D**.

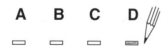

1. Which of the following oxides forms an aqueous solution with pH greater than 7?

   A Carbon dioxide

   B Copper(II) oxide

   C Sulphur dioxide

   D Sodium oxide

2. In which of the following reactions is a positive ion reduced?

   A Iodide $\longrightarrow$ iodine

   B Nickel(II) $\longrightarrow$ nickel(III)

   C Cobalt(III) $\longrightarrow$ cobalt(II)

   D Sulphate $\longrightarrow$ sulphite

3. Which of the following elements is most likely to have a covalent network structure?

Element	Melting point/°C	Boiling point/°C	Density/ $g\,cm^{-3}$	Conduction when solid
A	44	280	1·82	No
B	660	2467	2·70	Yes
C	1410	2355	2·33	No
D	114	184	4·93	No

4. Two identical samples of copper(II) carbonate were added to an excess of $1\,mol\,l^{-1}$ hydrochloric acid and $1\,mol\,l^{-1}$ sulphuric acid respectively.

   Which of the following would have been different for the two reactions?

   A The pH of the final solution

   B The volume of gas produced

   C The mass of water formed

   D The mass of copper(II) carbonate dissolved

5. The graph shows how the rate of a reaction varies with the concentration of one of the reactants.

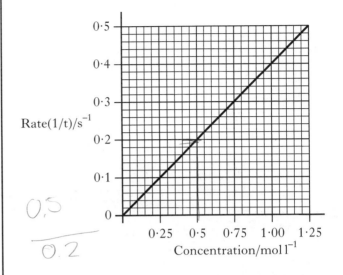

   What was the reaction time, in seconds, when the concentration of the reactant was $0·50\,mol\,l^{-1}$?

   A 0·2

   B 0·5

   C 2·0

   D 5·0

6. 10 g of magnesium is added to 1 litre of $1\,mol\,l^{-1}$ copper(II) sulphate solution and the mixture stirred until no further reaction occurs.

   Which of the following is a result of this reaction?

   A All the magnesium reacts.

   B 63·5 g of copper is displaced.

   C 2 mol of copper is displaced.

   D The resulting solution is colourless.

   [Turn over

7. A reaction was carried out with and without a catalyst as shown in the energy diagram.

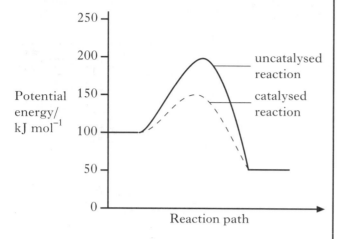

What is the enthalpy change, in kJ mol^{-1}, for the catalysed reaction?

A   −100

B   −50

C   +50

D   +100

8. Ethanol (C_2H_5OH) has a different enthalpy of combustion from dimethyl ether (CH_3OCH_3). This is because the compounds have different

A   boiling points

B   molecular masses

C   products of combustion

D   bonds within the molecules.

9. Which of the following compounds has the greatest ionic character?

A   Caesium fluoride

B   Caesium iodide

C   Sodium fluoride

D   Sodium iodide

10. Which line in the table is likely to be correct for the element francium?

	State at 30 °C	First ionisation energy/kJ mol^{-1}
A	solid	less than 382
B	liquid	less than 382
C	solid	greater than 382
D	liquid	greater than 382

11. Which of the following equations represents the first ionisation energy of fluorine?

A $F^-(g) \rightarrow F(g) + e^-$

B $F^-(g) \rightarrow \frac{1}{2}F_2(g) + e^-$

C $F(g) \rightarrow F^+(g) + e^-$

D $\frac{1}{2}F_2(g) \rightarrow F^+(g) + e^-$

12. The two hydrogen atoms in a molecule of hydrogen are held together by

A   a hydrogen bond

B   a polar covalent bond

C   a non-polar covalent bond

D   a van der Waals' force.

13. In which of the following compounds would hydrogen bonding **not** occur?

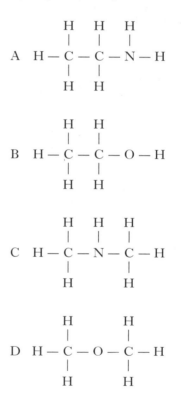

14. Which of the following shows the types of bonding in **decreasing** order of strength?

A   Covalent : hydrogen : van der Waals'

B   Covalent : van der Waals' : hydrogen

C   Hydrogen : covalent : van der Waals'

D   Van der Waals' : hydrogen : covalent

15. What type of bonding and structure is found in a fullerene?

    A Ionic lattice

    B Metallic lattice

    C Covalent network

    D Covalent molecular

16. Some covalent compounds are made up of molecules that contain polar bonds but the molecules are overall non-polar.

    Which of the following covalent compounds is made up of non-polar molecules?

    A Ammonia

    B Water

    C Carbon tetrachloride

    D Hydrogen fluoride

17. The Avogadro Constant is the same as the number of

    A ions in 1 mol of NaCl

    B atoms in 1 mol of hydrogen gas

    C electrons in 1 mol of helium gas

    D molecules in 1 mol of oxygen gas.

18. Which of the following gas samples has the same volume as 7 g of carbon monoxide?

    (All volumes are measured at the same temperature and pressure.)

    A 1 g of hydrogen

    B 3·5 g of nitrogen

    C 10 g of argon

    D 35·5 g of chlorine

19. What volume of oxygen (in litres) would be required for the complete combustion of a gaseous mixture containing 1 litre of carbon monoxide and 3 litres of hydrogen?

    (All volumes are measured at the same temperature and pressure.)

    A 1

    B 2

    C 3

    D 4

20. Which of the following pollutants, produced during internal combustion in a car engine, is **not** the result of incomplete combustion?

    A Carbon

    B Carbon monoxide

    C Hydrocarbons

    D Nitrogen dioxide

21. Which of the following compounds does **not** have isomeric structures?

    A C_2HCl_3

    B $C_2H_4Cl_2$

    C Propene

    D Propan-1-ol

22. Which of the following compounds is an alkanone?

    A $CH_3 - CH_2 - \overset{\displaystyle O}{\overset{\displaystyle \|}{C}} - H$

    B $CH_3 - \overset{\displaystyle O}{\overset{\displaystyle \|}{C}} - O - CH_3$

    C $CH_3 - \overset{\displaystyle O}{\overset{\displaystyle \|}{C}} - CH_3$

    D $CH_3 - \overset{\displaystyle O}{\overset{\displaystyle \|}{C}} - OH$

23. What organic compound is produced by the dehydration of ethanol?

    A Ethane

    B Ethene

    C Ethanal

    D Ethanoic acid

24. The production of synthesis gas from methane involves

    A steam reforming

    B catalytic cracking

    C hydration

    D oxidation.

25. Compound **X** reacted with hot copper(II) oxide and the organic product did not give a colour change when heated with Fehling's solution.

    Compound **X** could be

    A butan-1-ol

    B butan-2-ol

    C butanone

    D butanoic acid.

26. Part of a polymer is shown.

    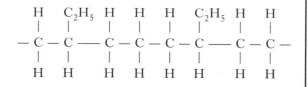

    Which two alkenes were used to make this polymer?

    A Ethene and propene

    B Ethene and but-1-ene

    C Propene and but-1-ene

    D Ethene and but-2-ene

27. Ammonia solution may be used to distinguish $Fe^{2+}(aq)$ from $Fe^{3+}(aq)$ as follows:

    $Fe^{2+}(aq)$ gives a green precipitate of $Fe(OH)_2$;

    $Fe^{3+}(aq)$ gives a brown precipitate of $Fe(OH)_3$.

    Which of the following types of compound is most likely to give similar results if used instead of ammonia?

    A An alcohol

    B An aldehyde

    C An amine

    D A carboxylic acid

28. Which of the following reactions takes place during the 'hardening' of vegetable oil?

    A Addition

    B Hydrolysis

    C Dehydration

    D Oxidation

29. Fats are formed by the condensation reaction between glycerol molecules and fatty acid molecules.

    In this reaction the mole ratio of glycerol molecules to fatty acid molecules is

    A 1 : 1

    B 1 : 2

    C 1 : 3

    D 3 : 1.

30. Which of the following graphs shows how the rate of reaction varies with temperature for the fermentation of glucose?

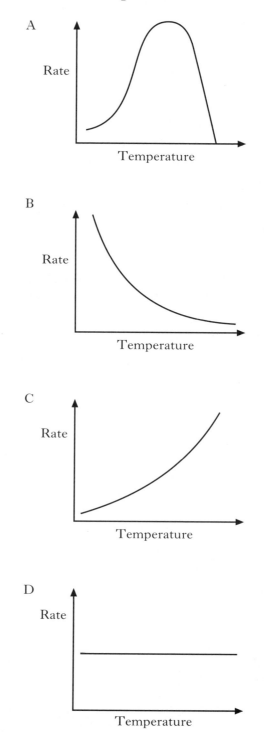

31. Which of the following is the best description of a feedstock?

A A consumer product such as a textile, plastic or detergent.

B A complex chemical that has been synthesised from small molecules.

C A mixture of chemicals formed by the cracking of the naphtha fraction from oil.

D A chemical from which other chemicals can be extracted or synthesised.

32.
$$S(s) \quad + \quad H_2(g) \quad \rightarrow \quad H_2S(g)$$
$$\Delta H = a$$
$$H_2(g) \quad + \quad \tfrac{1}{2}O_2(g) \quad \rightarrow \quad H_2O(\ell)$$
$$\Delta H = b$$
$$S(s) \quad + \quad O_2(g) \quad \rightarrow \quad SO_2(g)$$
$$\Delta H = c$$
$$H_2S(g) \quad + \quad 1\tfrac{1}{2}O_2(g) \quad \rightarrow \quad H_2O(\ell) \quad + \quad SO_2(g)$$
$$\Delta H = d$$

What is the relationship between a, b, c and d?

A $a = b + c - d$

B $a = d - b - c$

C $a = b - c - d$

D $a = d + c - b$

33. A catalyst is added to a reaction at equilibrium.

Which of the following does **not** apply?

A The rate of the forward reaction increases.

B The rate of the reverse reaction increases.

C The position of equilibrium remains unchanged.

D The position of equilibrium shifts to the right.

34. Steam and carbon monoxide react to form an equilibrium mixture.

$$CO(g) + H_2O(g) \rightleftharpoons H_2(g) + CO_2(g)$$

Which of the following graphs shows how the rates of the forward and reverse reactions change when carbon monoxide and steam are mixed?

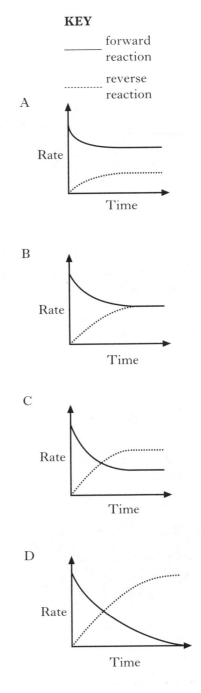

[Turn over

35. Solid sodium sulphite is dissolved in distilled water, producing an alkaline solution.

Which of the following processes is the most important in causing this change?

A   Sodium ions reacting with hydroxide ions

B   Hydrogen ions reacting with sulphite ions

C   Sodium ions reacting with sulphite ions

D   Hydrogen ions reacting with hydroxide ions

36. Which of the following salts dissolves in water to form an acidic solution?

A CH_3COONa

B Na_2SO_4

C KCl

D NH_4NO_3

37. Iodide ions can be oxidised using acidified potassium permanganate solution.

The equations are:

$2I^-(aq) \rightarrow I_2(aq) + 2e^-$

$MnO_4^-(aq) + 8H^+(aq) + 5e^- \rightarrow Mn^{2+}(aq) + 4H_2O(\ell)$

How many moles of iodide ions are oxidised by one mole of permanganate ions?

A $1 \cdot 0$

B $2 \cdot 0$

C $2 \cdot 5$

D $5 \cdot 0$

38. In the electrolysis of molten magnesium chloride, 1 mol of magnesium is deposited at the negative electrode by

A   96 500 coulombs

B   193 000 coulombs

C   1 mol of electrons

D   24·3 mol of electrons.

39. Alpha, beta and gamma radiation is passed from a source through an electric field onto a photographic plate.

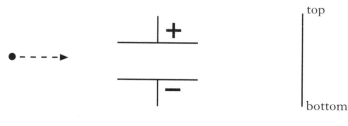

source of α, β electric field   photographic
and γ radiations   plate

Which of the following patterns will be produced on the photographic plate?

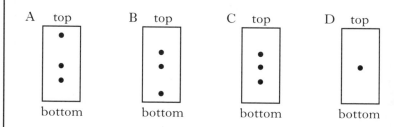

40. From which of the following could $^{32}_{15}P$ be produced by neutron capture?

A $^{33}_{15}P$

B $^{32}_{16}S$

C $^{31}_{15}P$

D $^{31}_{16}S$

Candidates are reminded that the answer sheet MUST be returned INSIDE the front cover of this answer book.

[Turn over for Section B on *Page ten*

Marks

SECTION B

All answers must be written clearly and legibly in ink.

1. (*a*) Lithium starts the second period of the Periodic Table.

Li	Be	B	C	N	O	F

What is the trend in electronegativity values across this period from Li to F?

increase.

1

(*b*) **Graph 1** shows the first four ionisation energies for aluminium.

Graph 1

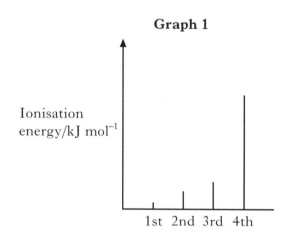

Why is the fourth ionisation energy of aluminium so much higher than the third ionisation energy?

*are more stable electron
arrangement, therefore more energy
is needed to remove*

1

Marks

1. (continued)

    (*c*) **Graph 2** shows the boiling points of the elements in Group 7 of the Periodic Table.

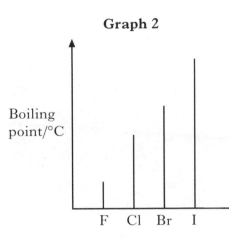

Graph 2

    Why do the boiling points increase down Group 7?

atoms increase in size, therefore,
increasing van der waals
force

1

(3)

[Turn over

Marks

2. Reactions are carried out in oil refineries to increase the octane number of petrol components. One such reaction is:

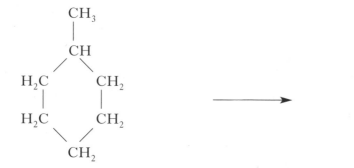

methylcyclohexane   methylbenzene

(a) The molecular formula for methylbenzene can be written C_xH_y.
Give the values for **x** and **y**.

x = 7

y = 8

1

(b) Hydrogen, the by-product of the reaction, can also be used as a fuel.
Give **one** advantage of using hydrogen as a fuel instead of petrol.

does not produce polluting gases

1

(c) In some countries oxygenates are added to the petrol. These improve the efficiency of burning. One common oxygenate is ethanol.
Give **one** other advantage of adding ethanol to petrol.

burns more efficiently and produces less CO than petrol

1

(3)

Marks

3. Alkanols can be oxidised to alkanoic acids.

$$CH_3CH_2CH_2OH \xrightarrow{\textbf{Step 1}} CH_3CH_2CHO \xrightarrow{\textbf{Step 2}} CH_3CH_2COOH$$

propan-1-ol   propanal   propanoic acid

(*a*) (i) Why can **Step 1** be described as an oxidation reaction?

1

(ii) Acidified potassium dichromate solution can be used to oxidise propanal in **Step 2**.

What colour change would be observed in this reaction?

orange - green

1

(*b*) Propan-1-ol and propanoic acid react to form an ester.

The mixture of excess reactants and ester product is poured onto sodium hydrogencarbonate solution.

(i) What evidence would show that an ester is formed?

smell.

1

(ii) Draw a structural formula for this ester.

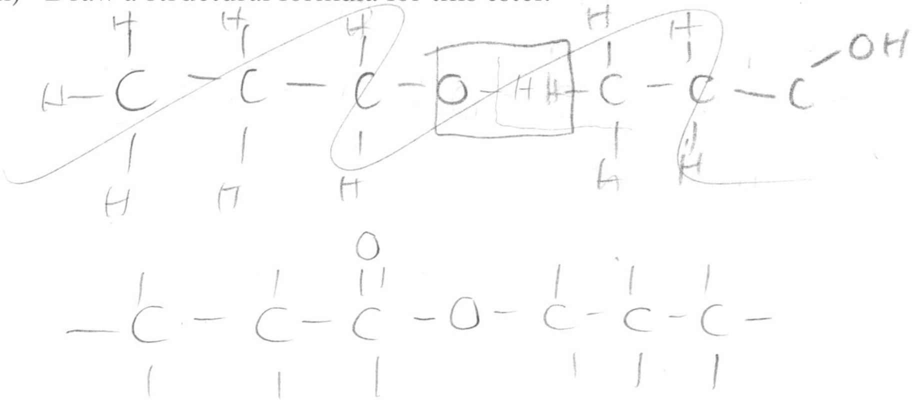

1

(4)

[*Turn over*

4. Ozone gas, $O_3(g)$, is made up of triatomic molecules.

Marks

(a) Ozone is present in the upper atmosphere.

Why is the ozone layer important for life on Earth?

It doesn't rans UV radiation

1

(b) The depletion of this layer is believed to be caused by chlorine radicals produced by the breakdown of certain CFCs. Some of the gaseous reactions are catalysed by ice crystals in clouds.

The crystals are acting as what type of catalyst?

heterogenous

1

(c) Ozone can be produced in the laboratory by electrical discharge.

$$3O_2(g) \rightarrow 2O_3(g)$$

Calculate the approximate number of $O_3(g)$ molecules produced from one mole of $O_2(g)$ molecules.

$$3O_2(g) \rightarrow 2O_3(g)$$

3mol — 2mol

x — 1mol

1

(3)

$3x = 2$

$x = \dfrac{2}{3}$

1mol — 6.02×10^{23}

$\dfrac{2}{3} \times x$

$x = 4.013$

DO NOT WRITE IN THIS MARGIN

Marks

5. Polymers can be classified as natural or synthetic.

(*a*) Keratin, a natural polymer, is a protein found in hair.

The hydrolysis of keratin produces different monomers of the type shown.

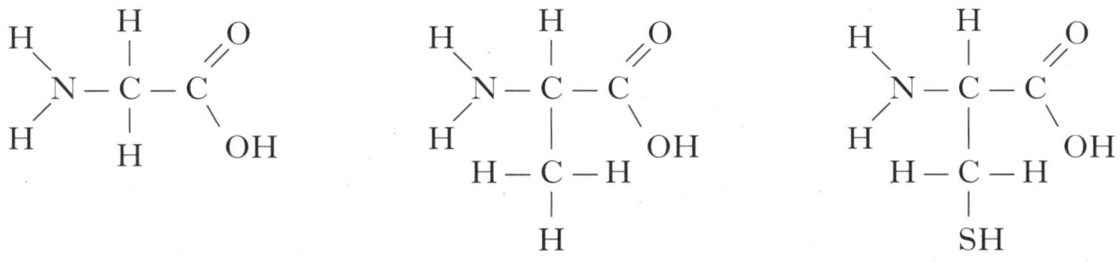

glycine alanine cysteine

(i) What name is given to monomers like glycine, alanine and cysteine?

amino acid

1

(ii) What is meant by a **hydrolysis** reaction?

being broken down by H_2O

1

(*b*) Dacron, a synthetic polymer, is used in heart surgery.

A section of the polymer is shown.

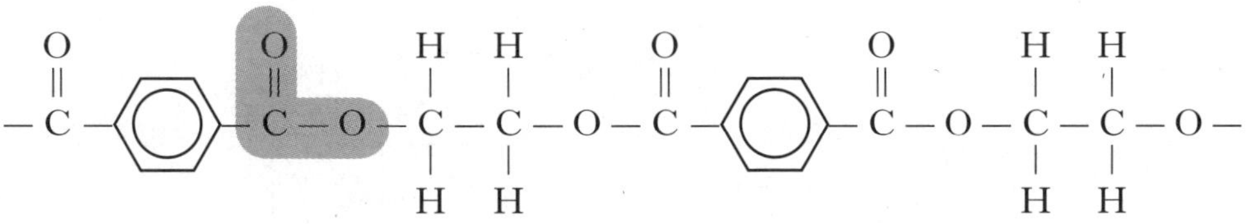

(i) What name is given to the link made by the shaded group of atoms in this section of the polymer?

ester

1

(ii) Why would this polymer be formed as a fibre and not as a resin?

no chance of cross links

1

(4)

Marks

6.   A student used the simple laboratory apparatus shown to determine the enthalpy of combustion of methanol.

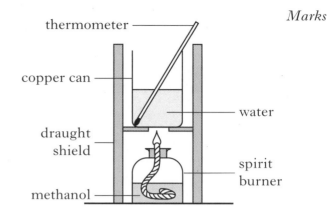

thermometer

copper can

water

draught shield

spirit burner

methanol

(*a*)   (i)   What measurements are needed to calculate the energy released by the burning methanol?

→ volume of water

→ change in temperature

1

(ii)   The student found that burning 0·370 g of methanol produces 3·86 kJ of energy.

Use this result to calculate the enthalpy of combustion of methanol.

-OH

methanol

CH₃OH

4 ┤ ├12
  └─┤4
    └16
= 32g

0.370g — 3.86kJ

32g — x

0.370x — 123.52

x — 333.8 kJ mol⁻¹

1

(*b*)   A more accurate value can be obtained using a bomb calorimeter.

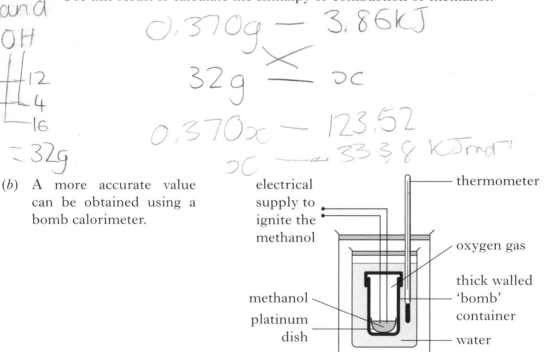

electrical supply to ignite the methanol

thermometer

oxygen gas

methanol

platinum dish

thick walled 'bomb' container

water

One reason for the more accurate value is that less heat is lost to the surroundings than in the simple laboratory method.

Give **one** other reason for the value being more accurate in the bomb calorimeter method.

complete combustion

1

(3)

Marks

7. An experiment was carried out to determine the rate of the reaction between hydrochloric acid and calcium carbonate chips. The rate of this reaction was followed by measuring the volume of gas released over a certain time.

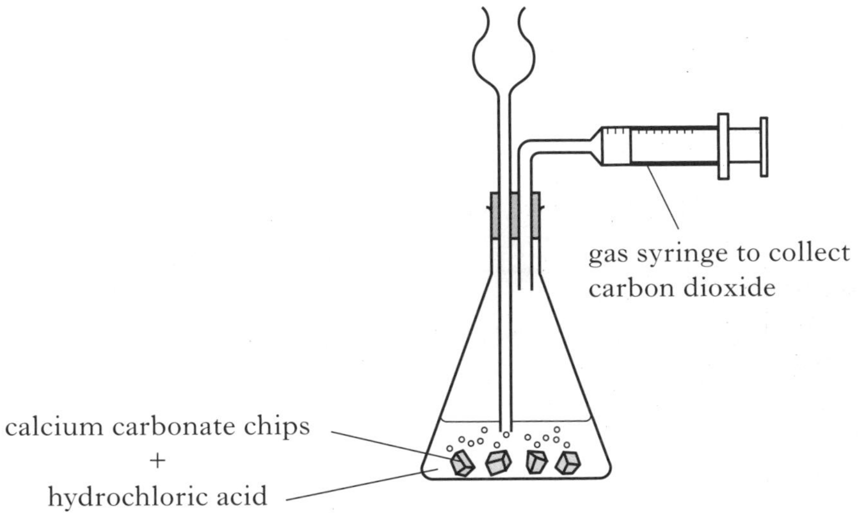

gas syringe to collect carbon dioxide

calcium carbonate chips
+
hydrochloric acid

(a) Describe a different way of measuring volume in order to follow the rate of this reaction.

1

(b) What other variable could be measured to follow the rate of this reaction?

concentration of acid

1

(2)

[Turn over

Marks

8. Ammonia is produced in industry by the Haber Process.

$$N_2(g) + 3H_2(g) \rightleftharpoons 2NH_3(g)$$

(a) State whether the industrial manufacture of ammonia is likely to be a batch or a continuous process.

continuous

1

(b) The graph shows how the percentage yield of ammonia changes with temperature at a pressure of 100 atmospheres.

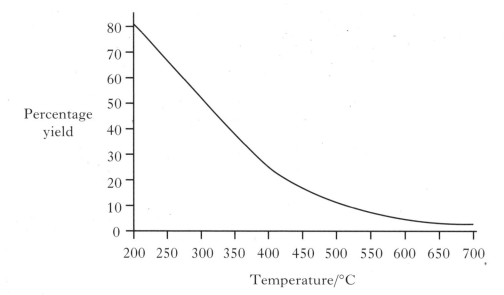

(i) A student correctly concludes from the graph that the production of ammonia is an exothermic process.

What is the reasoning that leads to this conclusion?

percentage yield decreases as temperature increases

1

(ii) **Explain clearly** why the industrial manufacture of ammonia is carried out at a pressure greater than 100 atmospheres.

High pressure - equilm to more the

2

Marks

8. (continued)

(c) Under certain conditions, 500 kg of nitrogen reacts with excess hydrogen to produce 405 kg of ammonia.

Calculate the <u>percentage yield</u> of ammonia under these conditions.

Show your working clearly.

$$N \quad\quad \overset{1 mol}{500 kg} - \overset{1 mol}{405 g}$$

$$L_{14g}$$

$$\frac{14 g}{500 kg} \times \frac{17 g}{x}$$

$$NH_3$$
$$L_3$$
$$L_{17g}$$

$$14x - 8500$$

$$x = 607.14$$

2

(6)

[Turn over

$$\% = \frac{A}{T} \times 100$$

$$= \frac{405}{607.14} \times 100$$

$$= 66.7\%$$

Marks

9. Primary, secondary and tertiary alkanols can be prepared by the reaction of carbonyl compounds with Grignard reagents.

Step 1

The Grignard reagent reacts with the carbonyl compound.

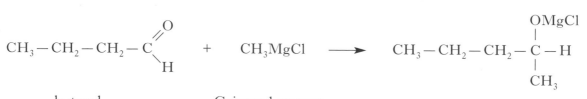

butanal   Grignard reagent

Step 2

The reaction of the product of **Step 1** with dilute acid produces the alkanol.

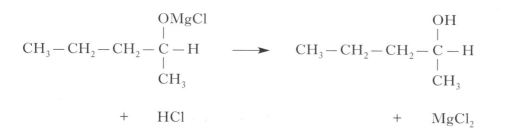

+   HCl   + $MgCl_2$

(a) Describe the difference between a primary, a secondary and a tertiary alkanol. You may wish to include labelled structures in your answer.

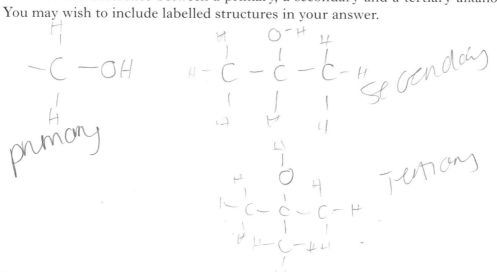

1

(b) Suggest a name for the type of reaction that takes place in **Step 1**.

1

Marks

9. (continued)

(*c*) The same Grignard reagent can be used to produce the alkanol below.

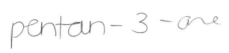

$$CH_3-CH_2-\underset{\underset{CH_3}{|}}{\overset{\overset{OH}{|}}{C}}-CH_2-CH_3$$

Name the carbonyl compound used in this reaction.

pentan – 3 – one

1
(3)

[Turn over

Marks

10. Sherbet contains a mixture of sodium hydrogencarbonate and tartaric acid. The fizzing sensation in the mouth is due to the carbon dioxide produced in the following reaction.

$$2NaHCO_3 \quad + \quad C_4H_6O_6 \quad \rightarrow \quad Na_2(C_4H_4O_6) \quad + \quad 2H_2O \quad + \quad 2CO_2$$

sodium   tartaric acid   sodium tartrate
hydrogencarbonate

(a) Name the type of reaction taking place.

neutralisation

1

(b) The chemical name for tartaric acid is 2,3-dihydroxybutanedioic acid.

Draw a structural formula for tartaric acid.

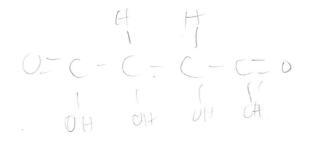

1

(c) In an experiment, a student found that adding water to 20 sherbet sweets produced $105 \, cm^3$ of carbon dioxide.

Assuming that sodium hydrogencarbonate is in excess, calculate the average mass of tartaric acid, in grams, in one sweet.

(Take the molar volume of carbon dioxide to be 24 litre mol^{-1}.)

Show your working clearly.

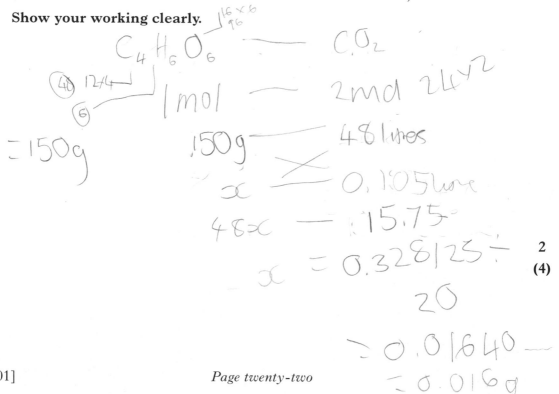

2
(4)

Marks

11. The following answers were taken from a student's examination paper.

The two answers are incorrect.

For each question, give the correct explanation.

(*a*) **Question**   As a rough guide, the rate of a reaction tends to double for every $10\,°C$ rise in temperature.

Why does a small increase in temperature produce a large increase in reaction rate?

Student answer   Because rising temperature increases the activation energy which increases the number of collisions which speeds up the reaction greatly.

Correct explanation   *more collision great or a equal to more answer*

1

(*b*) **Question**   Explain the difference in atomic size between potassium and chlorine atoms.

Student answer   A potassium nucleus has 19 protons but a chlorine nucleus has only 17 protons. The greater pull on the outer electron in the potassium atom means the atomic size of potassium is less than that of chlorine.

Correct explanation   *potassium is fute away from nucleus*

1

(2)

[Turn over

12. Barium hydroxide solution neutralises dilute sulphuric acid. A white precipitate of barium sulphate is formed in the reaction.

$$H_2SO_4(aq) \quad + \quad Ba(OH)_2(aq) \quad \rightarrow \quad BaSO_4(s) \quad + \quad 2H_2O(\ell)$$

The reaction can be followed by measuring the conductivity of the solution as barium hydroxide solution is added to dilute sulphuric acid.

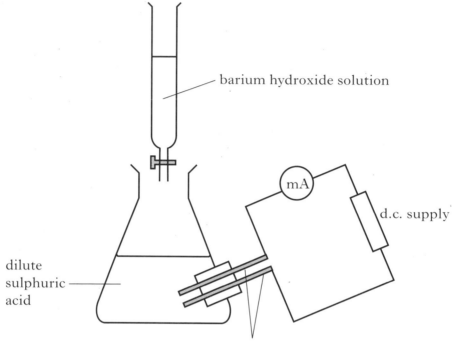

The graph shows how the conductivity changed in one experiment when barium hydroxide solution was added to $50\,cm^3$ of $0.01\,mol\,l^{-1}$ sulphuric acid.

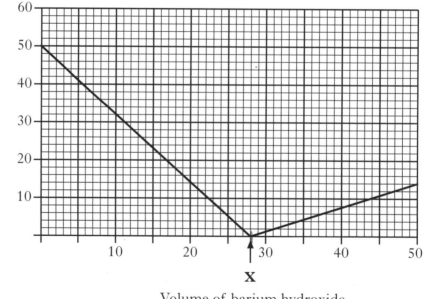

Marks

12. (continued)

(a) Point **X** corresponds to the end-point of the reaction.

Use the information on the graph to calculate the concentration of the barium hydroxide solution, in $mol\,l^{-1}$.

no moles $= V \times C$

$= 0.05 \times 0.01$

$= 0.005$

$28\,cm^3$ — 0.0005

1

(b) **Explain clearly** why the conductivity is very close to zero at the end-point.

Ions are not free to move and water contains few ions

2

(3)

[Turn over

DO NOT
WRITE IN
THIS
MARGIN

Marks

13. An electrolysis experiment was set up with two cells as shown.

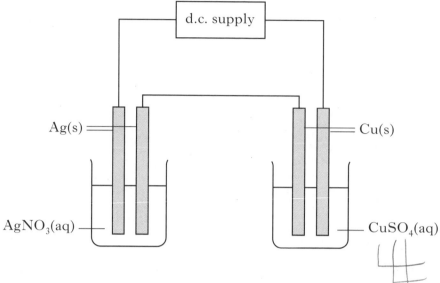

The reaction at the silver electrode is:

$$Ag^+(aq) \quad + \quad e^- \quad \rightarrow \quad Ag(s)$$

The mass of silver deposited in the reaction was 0·365 g.

(a) In addition to measuring the time, what **two** changes to the circuit would need to be made to find accurately the quantity of charge required to deposit 0·365 g of silver?

add a variable resistor

ammeter.

1

(b) What mass of copper would be deposited in the same time?

Show your working clearly.

2

(3)

14. (*a*) A student compared the properties of equal concentrations of aqueous solutions of hydrochloric acid and ethanoic acid.

Experiment		Hydrochloric acid	Ethanoic acid
1	Rate of reaction with magnesium	fast	faster / (slower) / same
2	Electrical conductivity	80 mA	higher / (lower) / same
3	Volume of $0 \cdot 1 \, mol \, l^{-1}$ sodium hydroxide to neutralise 20 cm^3 acid	20 cm^3	more / less / (same)

The result for ethanoic acid has been circled for experiment 1.

Circle the expected results for ethanoic acid in experiments 2 and 3.

1

(*b*) Some of the hydrogen atoms in ethanoic acid can be replaced by chlorine atoms to give three chloroethanoic acids. The student measured the pH of aqueous solutions of the four related acids. All the acids had the same concentration.

Acid name	Molecular formula	pH
ethanoic acid	CH_3COOH	$2 \cdot 3$
chloroethanoic acid	$CH_2ClCOOH$	$1 \cdot 4$
dichloroethanoic acid	$CHCl_2COOH$	$1 \cdot 0$
trichloroethanoic acid	CCl_3COOH	$0 \cdot 2$

(i) What is the concentration of hydroxide ions, in mol l^{-1}, in the dichloroethanoic acid solution?

pHO.1

$-.H^+ = 1 \times 10^{-1}$ $= 1 \times 10^{-14}$

$OH^- = 1 \times 10^{-13} \, mol \, l^{-1}$

1

(ii) **Explain clearly** the way in which the number of chlorine atoms in the acid molecules affects the strength of the acid.

2

(4)

Marks

15. (*a*) Methane is produced in the reaction of aluminium carbide with water.

$$Al_4C_3 \quad + \quad H_2O \quad \rightarrow \quad Al(OH)_3 \quad + \quad CH_4$$

Balance the above equation.

1

(*b*) Silane, silicon hydride, is formed in the reaction of silicon with hydrogen.

$$Si(s) \quad + \quad 2H_2(g) \quad \rightarrow \quad SiH_4(g)$$
$$\text{silane}$$

The enthalpy change for this reaction is called the enthalpy of formation of silane.

The combustion of silane gives silicon dioxide and water.

$$SiH_4(g) \quad + \quad 2O_2(g) \quad \rightarrow \quad SiO_2(s) \quad + \quad 2H_2O(\ell) \quad \Delta H = -1517\,kJ\,mol^{-1}$$

The enthalpy of combustion of silicon is $-911\,kJ\,mol^{-1}$.

Use this information and the enthalpy of combustion of hydrogen in the data booklet to calculate the enthalpy of formation of silane, in $kJ\,mol^{-1}$.

Show your working clearly.

$$\Delta H - 911$$

2

(3)

DO NOT WRITE IN THIS MARGIN

Marks

16. Thorium-227 decays by alpha emission.

(*a*) Complete the nuclear equation for the alpha decay of thorium-227.

^{227}Th $\rightarrow$ $^{4}_{2}$He 223

1

(*b*) A sample of thorium-227 was placed in a wooden box. A radiation detector was held 10 cm away from the box.

Why was alpha radiation not detected?

Aphas wavelength is extremly short and will only travel a short distance. Also it can not penetrate ...

1

(*c*) Thorium-227 has a half-life of 19 days. If 0·42 g of thorium-227 has decayed after 57 days, calculate the initial mass of thorium-227, in grams.

57 ÷ 19 = 3 half lifes

0.42 × 2

0.42 — 0.84 —

1
(3)

[Turn over

Marks

17. Carbon-13 NMR is a technique used in chemistry to determine the structure of organic compounds.

   (a) Calculate the neutron to proton ratio in an atom of carbon-13.

   7 : 6

   1

   (b) The technique allows a carbon atom in a molecule to be identified by its 'chemical shift'. This value depends on the other atoms bonded to the carbon atom.

   Shift table

Carbon environment	Chemical shift/ppm
$C = O$ (in ketones)	205 – 220
$C = O$ (in aldehydes)	190 – 205
$C = O$ (in acids and esters)	170 – 185
$C = C$ (in alkenes)	115 – 140
$C \equiv C$ (in alkynes)	70 – 95
$-CH$	25 – 50
$-CH_2$	16 – 40
$-CH_3$	5 – 15

   In a carbon-13 NMR spectrum, the number of lines correspond to the number of chemically different carbon atoms and the position of the line (the value of the chemical shift) indicates the type of carbon atom.

Marks

17. (*b*) (continued)

The spectrum for propanal is shown.

Spectrum 1

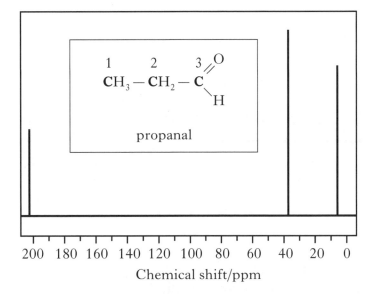

200 180 160 140 120 100 80 60 40 20 0
Chemical shift/ppm

(i) Use the table of chemical shifts to label each of the peaks on the spectrum with a number to match the carbon atom in propanal that is responsible for the peak.

1

(ii) Hydrocarbon **X** has a relative formula mass of 54. Hydrocarbon **X** reacts with hydrogen. One of the products, hydrocarbon **Y**, has a relative formula mass of 56.

The carbon-13 NMR spectrum for hydrocarbon **Y** is shown below.

Spectrum 2

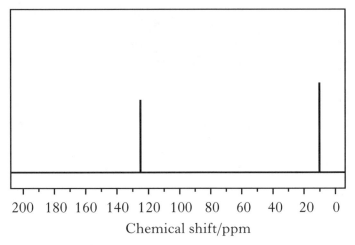

200 180 160 140 120 100 80 60 40 20 0
Chemical shift/ppm

Name hydrocarbon **Y**.

1

(3)

Marks

18. The number of moles of carbon monoxide in a sample of air can be measured as follows.

    Step 1   The carbon monoxide reacts with iodine(V) oxide, producing iodine.

    $$5CO(g) \quad + \quad I_2O_5(s) \quad \rightarrow \quad I_2(s) \quad + \quad 5CO_2(g)$$

    Step 2   The iodine is then dissolved in potassium iodide solution and titrated against sodium thiosulphate solution.

    $$I_2(aq) \quad + \quad 2S_2O_3^{2-}(aq) \quad \rightarrow \quad S_4O_6^{2-}(aq) \quad + \quad 2I^-(aq)$$

    (a) Write the ion-electron equation for the oxidation reaction in **Step 2**.

    1

    (b) Name a chemical that can be used to indicate when all of the iodine has been removed in the reaction taking place in **Step 2**.

    1

    (c) If $50.4 \, cm^3$ of $0.10 \, mol \, l^{-1}$ sodium thiosulphate solution was used in a titration, calculate the number of moles of carbon monoxide in the sample of air.

    Show your working clearly.

    2
    (4)

[END OF QUESTION PAPER]

ADDITIONAL SPACE FOR ANSWERS

ADDITIONAL SPACE FOR ANSWERS

ADDITIONAL SPACE FOR ANSWERS

[BLANK PAGE]

HIGHER

2010

[BLANK PAGE]

FOR OFFICIAL USE

Total
Section B

X012/301

NATIONAL
QUALIFICATIONS
2010

WEDNESDAY, 2 JUNE
9.00 AM – 11.30 AM

CHEMISTRY
HIGHER

Fill in these boxes and read what is printed below.

Full name of centre

Town

Forename(s)

Surname

Date of birth

Day	Month	Year	Scottish candidate number	Number of seat

Reference may be made to the Chemistry Higher and Advanced Higher Data Booklet.

SECTION A   Questions 1 40 (40 marks)

Instructions for completion of **Section A** are given on page two.

For this section of the examination you must use an **HB pencil**.

SECTION B (60 marks)

1   All questions should be attempted.

2   The questions may be answered in any order but all answers are to be written in the spaces provided in this answer book, **and must be written clearly and legibly in ink**.

3   Rough work, if any should be necessary, should be written in this book and then scored through when the fair copy has been written. If further space is required, a supplementary sheet for rough work may be obtained from the Invigilator.

4   Additional space for answers will be found at the end of the book. If further space is required, supplementary sheets may be obtained from the Invigilator and should be inserted inside the **front** cover of this book.

5   The size of the space provided for an answer should not be taken as an indication of how much to write. It is not necessary to use all the space.

6   Before leaving the examination room you must give this book to the Invigilator. If you do not, you may lose all the marks for this paper.

SECTION A

Read carefully

1 Check that the answer sheet provided is for **Chemistry Higher (Section A)**.

2 For this section of the examination you must use an **HB pencil** and, where necessary, an eraser.

3 Check that the answer sheet you have been given has **your name**, **date of birth**, **SCN** (Scottish Candidate Number) and **Centre Name** printed on it.

   Do not change any of these details.

4 If any of this information is wrong, tell the Invigilator immediately.

5 If this information is correct, **print** your name and seat number in the boxes provided.

6 The answer to each question is **either** A, B, C or D. Decide what your answer is, then, using your pencil, put a horizontal line in the space provided (see sample question below).

7 There is **only one correct** answer to each question.

8 Any rough working should be done on the question paper or the rough working sheet, **not** on your answer sheet.

9 At the end of the examination, put the **answer sheet for Section A inside the front cover of your answer book**.

Sample Question

To show that the ink in a ball-pen consists of a mixture of dyes, the method of separation would be

   A chromatography

   B fractional distillation

   C fractional crystallisation

   D filtration.

The correct answer is **A**—chromatography. The answer **A** has been clearly marked in **pencil** with a horizontal line (see below).

Changing an answer

If you decide to change your answer, carefully erase your first answer and using your pencil, fill in the answer you want. The answer below has been changed to **D**.

1. Which of the following gases would dissolve in water to form an alkali?

   A   HBr

   B NH_3

   C CO_2

   D CH_4

2. Which of the following pairs of solutions is most likely to produce a precipitate when mixed?

   A   Magnesium nitrate + sodium chloride

   B   Magnesium nitrate + sodium sulphate

   C   Silver nitrate + sodium chloride

   D   Silver nitrate + sodium sulphate

3. 0·5 mol of copper(II) chloride and 0·5 mol of copper(II) sulphate are dissolved together in water and made up to 500 cm^3 of solution.

   What is the concentration of Cu^{2+}(aq) ions in the solution in $mol\,l^{-1}$?

   A   0·5

   B   1·0

   C   2·0

   D   4·0

4. For any chemical, its temperature is a measure of

   A   the average kinetic energy of the particles that react

   B   the average kinetic energy of all the particles

   C   the activation energy

   D   the minimum kinetic energy required before reaction occurs.

5. 1 mol of hydrogen gas and 1 mol of iodine vapour were mixed and allowed to react. After *t* seconds, 0·8 mol of hydrogen remained.

   The number of moles of hydrogen iodide formed at *t* seconds was

   A   0·2

   B   0·4

   C   0·8

   D   1·6.

6. Excess zinc was added to 100 cm^3 of hydrochloric acid, concentration 1 $mol\,l^{-1}$.

   Graph I refers to this reaction.

   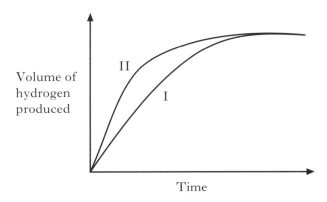

   Graph II could be for

   A   excess zinc reacting with 100 cm^3 of hydrochloric acid, concentration 2 $mol\,l^{-1}$

   B   excess zinc reacting with 100 cm^3 of sulphuric acid, concentration 1 $mol\,l^{-1}$

   C   excess zinc reacting with 100 cm^3 of ethanoic acid, concentration 1 $mol\,l^{-1}$

   D   excess magnesium reacting with 100 cm^3 of hydrochloric acid, concentration 1 $mol\,l^{-1}$.

7. Which of the following is **not** a correct statement about the effect of a catalyst?

   The catalyst

   A   provides an alternative route to the products

   B   lowers the energy that molecules need for successful collisions

   C   provides energy so that more molecules have successful collisions

   D   forms bonds with reacting molecules.

8. A potential energy diagram can be used to show the activation energy (E_A) and the enthalpy change (ΔH) for a reaction.

   Which of the following combinations of E_A and ΔH could **never** be obtained for a reaction?

   A $E_A = 50\,kJ\,mol^{-1}$ and $\Delta H = -100\,kJ\,mol^{-1}$

   B $E_A = 50\,kJ\,mol^{-1}$ and $\Delta H = +100\,kJ\,mol^{-1}$

   C $E_A = 100\,kJ\,mol^{-1}$ and $\Delta H = +50\,kJ\,mol^{-1}$

   D $E_A = 100\,kJ\,mol^{-1}$ and $\Delta H = -50\,kJ\,mol^{-1}$

[Turn over

9. As the relative atomic mass in the halogens increases

   A   the boiling point increases

   B   the density decreases

   C   the first ionisation energy increases

   D   the atomic size decreases.

10. The table shows the first three ionisation energies of aluminium.

Ionisation energy/kJ mol^{-1}		
1st	2nd	3rd
584	1830	2760

Using this information, what is the enthalpy change, in kJ mol^{-1}, for the following reaction?

$$Al^{3+}(g) + 2e^- \rightarrow Al^+(g)$$

   A   +2176

   B   −2176

   C   +4590

   D   −4590

11. When two atoms form a non-polar covalent bond, the two atoms **must** have

   A   the same atomic size

   B   the same electronegativity

   C   the same ionisation energy

   D   the same number of outer electrons.

12. In which of the following liquids does hydrogen bonding occur?

   A   Ethanoic acid

   B   Ethyl ethanoate

   C   Hexane

   D   Hex-1-ene

13. Which line in the table shows the correct entries for tetrafluoroethene?

	Polar bonds?	Polar molecules?
A	yes	yes
B	yes	no
C	no	no
D	no	yes

14. Element **X** was found to have the following properties.

   (i)   It does not conduct electricity when solid.

   (ii)   It forms a gaseous oxide.

   (iii)   It is a solid at room temperature.

Element **X** could be

   A   magnesium

   B   silicon

   C   nitrogen

   D   sulphur.

15. The Avogadro Constant is the same as the number of

   A   molecules in 16 g of oxygen

   B   ions in 1 litre of sodium chloride solution, concentration 1 mol l^{-1}

   C   atoms in 24 g of carbon

   D   molecules in 2 g of hydrogen.

16. Which of the following contains one mole of neutrons?

   A   1 g of $^{1}_{1}H$

   B   1 g of $^{12}_{6}C$

   C   2 g of $^{24}_{12}Mg$

   D   2 g of $^{22}_{10}Ne$

17. $20\,cm^3$ of ammonia gas reacted with an excess of heated copper(II) oxide.

$$3CuO + 2NH_3 \rightarrow 3Cu + 3H_2O + N_2$$

Assuming all measurements were made at $200\,°C$, what would be the volume of gaseous products?

A $10\,cm^3$

B $20\,cm^3$

C $30\,cm^3$

D $40\,cm^3$

18. Which of the following fuels can be produced by the fermentation of biological material under anaerobic conditions?

A Methane

B Ethane

C Propane

D Butane

19. Rum flavouring is based on the compound with the formula shown.

$$CH_3CH_2CH_2C \overset{\displaystyle O}{\underset{\displaystyle OCH_2CH_3}{}}$$

It can be made from

A ethanol and butanoic acid

B propanol and ethanoic acid

C butanol and methanoic acid

D propanol and propanoic acid.

20. Which of the following structural formulae represents a tertiary alcohol?

A $CH_3 - \underset{\underset{\displaystyle CH_3}{|}}{\overset{\overset{\displaystyle CH_3}{|}}{C}} - CH_2 - OH$

B $CH_3 - \underset{\underset{\displaystyle OH}{|}}{\overset{\overset{\displaystyle CH_3}{|}}{C}} - CH_2 - CH_3$

C $CH_3 - CH_2 - CH_2 - \underset{\underset{\displaystyle OH}{|}}{\overset{\overset{\displaystyle H}{|}}{C}} - CH_3$

D $CH_3 - CH_2 - \underset{\underset{\displaystyle OH}{|}}{\overset{\overset{\displaystyle H}{|}}{C}} - CH_2 - CH_3$

21. What is the product when one mole of chlorine gas reacts with one mole of ethyne?

A 1,1-Dichloroethene

B 1,1-Dichloroethane

C 1,2-Dichloroethene

D 1,2-Dichloroethane

[Turn over

22.

$$CH_3 - CH_2 - C\overset{O}{\underset{H}{\diagdown}}$$

Reaction **X** $\downarrow$

$$CH_3 - CH_2 - CH_2 - OH$$

Reaction **Y** $\downarrow$

$$CH_3 - CH = CH_2$$

Which line in the table correctly describes reactions **X** and **Y**?

	Reaction X	Reaction Y
A	oxidation	dehydration
B	oxidation	condensation
C	reduction	dehydration
D	reduction	condensation

23. Ozone has an important role in the upper atmosphere because it

A absorbs ultraviolet radiation

B absorbs certain CFCs

C reflects ultraviolet radiation

D reflects certain CFCs.

24. Synthesis gas consists mainly of

A CH_4 alone

B CH_4 and CO

C CO and H_2

D CH_4, CO and H_2.

25. Ethene is used in the manufacture of addition polymers.

What type of reaction is used to produce ethene from ethane?

A Cracking

B Addition

C Oxidation

D Hydrogenation

26. Polyester fibres and cured polyester resins are both very strong.

Which line in the table correctly describes the structure of these polyesters?

	Fibre	Cured resin
A	cross-linked	cross-linked
B	linear	linear
C	cross-linked	linear
D	linear	cross-linked

27. Part of a polymer chain is shown below.

$$-O-\overset{O}{\overset{\|}{C}}-(CH_2)_4-\overset{O}{\overset{\|}{C}}-O-(CH_2)_6-O-\overset{O}{\overset{\|}{C}}-(CH_2)_4-\overset{O}{\overset{\|}{C}}-O-(CH_2)_6-O-$$

Which of the following compounds, when added to the reactants during polymerisation, would stop the polymer chain from getting too long?

A $HO-\overset{O}{\overset{\|}{C}}-(CH_2)_4-\overset{O}{\overset{\|}{C}}-OH$

B $HO-(CH_2)_6-OH$

C $HO-(CH_2)_5-\overset{O}{\overset{\|}{C}}-OH$

D $CH_3-(CH_2)_4-CH_2-OH$

28. Which of the following fatty acids is unsaturated?

A $C_{19}H_{39}COOH$

B $C_{21}H_{43}COOH$

C $C_{17}H_{31}COOH$

D $C_{13}H_{27}COOH$

29. Which of the following alcohols is likely to be obtained on hydrolysis of butter?

A $CH_3 - CH_2 - CH_2 - OH$

B $CH_3 - CH - CH_3$
$\qquad \quad |$
$\qquad \quad OH$

C $CH_2 - OH$
$\quad \; |$
$\; \; CH_2$
$\quad \; |$
$\; \; CH_2 - OH$

D $CH_2 - OH$
$\quad \; |$
$\; \; CH - OH$
$\quad \; |$
$\; \; CH_2 - OH$

30. Amino acids are converted into proteins by

A hydration

B hydrolysis

C hydrogenation

D condensation.

31. Which of the following compounds is a raw material in the chemical industry?

A Ammonia

B Calcium carbonate

C Hexane

D Nitric acid

32. Given the equations

$Mg(s) + 2H^+(aq) \rightarrow Mg^{2+}(aq) + H_2(g)$
$\qquad \qquad \qquad \qquad \qquad \Delta H = a \, J \, mol^{-1}$

$Zn(s) + 2H^+(aq) \rightarrow Zn^{2+}(aq) + H_2(g)$
$\qquad \qquad \qquad \qquad \qquad \Delta H = b \, J \, mol^{-1}$

$Mg(s) + Zn^{2+}(aq) \rightarrow Mg^{2+}(aq) + Zn(s)$
$\qquad \qquad \qquad \qquad \qquad \Delta H = c \, J \, mol^{-1}$

then, according to Hess's Law

A $c = a - b$

B $c = a + b$

C $c = b - a$

D $c = -b - a.$

33. In which of the following reactions would an increase in pressure cause the equilibrium position to move to the left?

A $CO(g) + H_2O(g) \rightleftharpoons CO_2(g) + H_2(g)$

B $CH_4(g) + H_2O(g) \rightleftharpoons CO(g) + 3H_2(g)$

C $Fe_2O_3(s) + 3CO(g) \rightleftharpoons 2Fe(s) + 3CO_2(g)$

D $N_2(g) + 3H_2(g) \rightleftharpoons 2NH_3(g)$

34. If ammonia is added to a solution containing copper(II) ions an equilibrium is set up.

$Cu^{2+}(aq) + 2OH^-(aq) + 4NH_3(aq) \rightleftharpoons Cu(NH_3)_4(OH)_2(aq)$
$\qquad \qquad \qquad \qquad \qquad \qquad \qquad \text{(deep blue)}$

If acid is added to this equilibrium system

A the intensity of the deep blue colour will increase

B the equilibrium position will move to the right

C the concentration of $Cu^{2+}(aq)$ ions will increase

D the equilibrium position will not be affected.

35. Which of the following is the best description of a $0.1 \, mol \, l^{-1}$ solution of hydrochloric acid?

A Dilute solution of a weak acid

B Dilute solution of a strong acid

C Concentrated solution of a weak acid

D Concentrated solution of a strong acid

[Turn over

36. A solution has a negative pH value.

This solution

A   neutralises $H^+(aq)$ ions

B   contains no $OH^-(aq)$ ions

C   has a high concentration of $H^+(aq)$ ions

D   contains neither $H^+(aq)$ ions nor $OH^-(aq)$ ions.

37. When a certain aqueous solution is diluted, its conductivity decreases but its pH remains constant.

It could be

A   ethanoic acid

B   sodium chloride

C   sodium hydroxide

D   nitric acid.

38. Equal volumes of four $1 \, mol \, l^{-1}$ solutions were compared.

Which of the following $1 \, mol \, l^{-1}$ solutions contains the most ions?

A   Nitric acid

B   Hydrochloric acid

C   Ethanoic acid

D   Sulphuric acid

39. In which reaction is hydrogen gas acting as an oxidising agent?

A $H_2 + CuO \rightarrow H_2O + Cu$

B $H_2 + C_2H_4 \rightarrow C_2H_6$

C $H_2 + Cl_2 \rightarrow 2HCl$

D $H_2 + 2Na \rightarrow 2NaH$

40. Which particle will be formed when an atom of $^{211}_{83}Bi$ emits an α-particle and the decay product then emits a β-particle?

A $^{207}_{82}Pb$

B $^{208}_{81}Tl$

C $^{209}_{80}Hg$

D $^{210}_{79}Au$

Candidates are reminded that the answer sheet MUST be returned INSIDE the front cover of this answer book.

Marks

SECTION B

All answers must be written clearly and legibly in ink.

1. The elements lithium, boron and nitrogen are in the second period of the Periodic Table.

   Complete the table below to show **both** the bonding and structure of these three elements at room temperature.

Name of element	Bonding	Structure
lithium		lattice
boron		
nitrogen	covalent	

2

(2)

[Turn over

Marks

2. (*a*) Polyhydroxyamide is a recently developed fire-resistant polymer.

The monomers used to produce the polymer are shown.

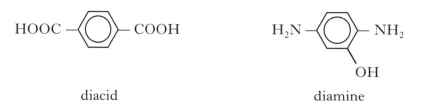

diacid   diamine

(i) How many hydrogen atoms are present in a molecule of the diamine molecule?

1

(ii) Draw a section of polyhydroxyamide showing **one** molecule of each monomer joined together.

1

(*b*) Poly(ethenol), another recently developed polymer, has an unusual property for a plastic.

What is this unusual property?

1

(3)

DO NOT WRITE IN THIS MARGIN

Marks

3. Atmospheric oxygen, $O_2(g)$, dissolves in the Earth's oceans forming dissolved oxygen, $O_2(aq)$, which is essential for aquatic life.

An equilibrium is established.

$$O_2(g) \quad + \quad (aq) \quad \rightleftharpoons \quad O_2(aq) \qquad \Delta H = -12{\cdot}1\ \text{kJ mol}^{-1}$$

(*a*) (i) What is meant by a reaction at "equilibrium"? **1**

(ii) What would happen to the concentration of dissolved oxygen if the temperature of the Earth's oceans increased? **1**

(*b*) A sample of oceanic water was found to contain 0·010 g of dissolved oxygen.

Calculate the number of moles of dissolved oxygen present in the sample. **1**

(3)

Marks

4. In the Hall-Heroult Process, aluminium is produced by the electrolysis of an ore containing aluminium oxide.

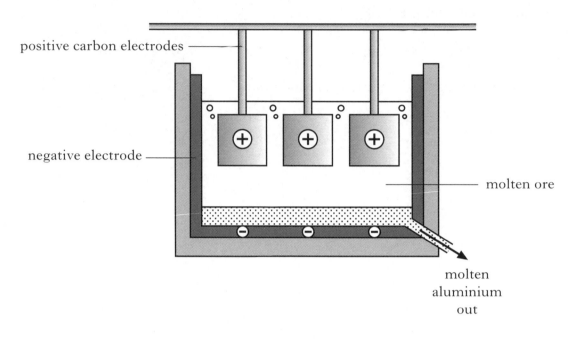

positive carbon electrodes

negative electrode

molten ore

molten aluminium out

(*a*) Suggest why the positive carbon electrodes need to be replaced regularly.

1

(*b*) Calculate the mass of aluminium, in grams, produced in 20 minutes when a current of 50 000 A is used.

Show your working clearly.

3

(4)

Marks

5. The reaction of oxalic acid with an acidified solution of potassium permanganate was studied to determine the effect of temperature changes on reaction rate.

$$5(COOH)_2(aq) + 6H^+(aq) + 2MnO_4^-(aq) \rightarrow 2Mn^{2+}(aq) + 10CO_2(g) + 8H_2O(\ell)$$

The reaction was carried out at several temperatures between 40 °C and 60 °C. The end of the reaction was indicated by a colour change from purple to colourless.

(*a*)   (i) State **two** factors that should be kept the same in these experiments.

1

(ii) Why is it difficult to measure an accurate value for the reaction time when the reaction is carried out at room temperature?

1

(*b*) Sketch a graph to show how the rate varied with increasing temperature.

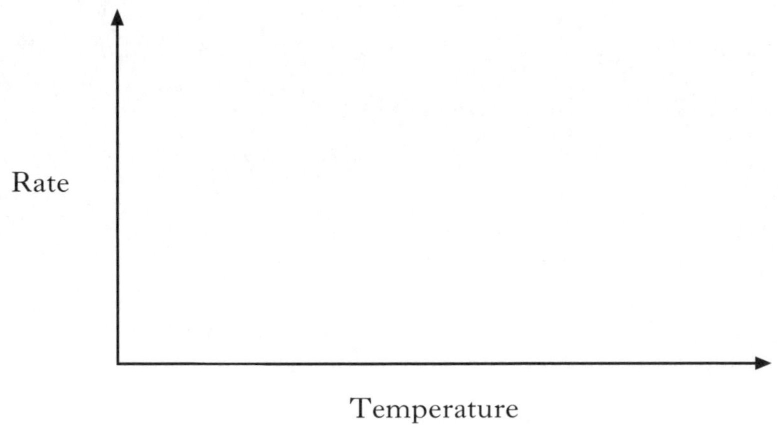

Rate

Temperature

1

(3)

[Turn over

Marks

6. Positron emission tomography, PET, is a technique that provides information about biochemical processes in the body.

Carbon-11, ^{11}C, is a positron-emitting radioisotope that is injected into the bloodstream.

A positron can be represented as $^{0}_{1}$e.

(a) Complete the nuclear equation for the decay of ^{11}C by positron-emission.

$$^{11}C \longrightarrow$$

1

(b) A sample of ^{11}C had an initial count rate of 640 counts min^{-1}. After 1 hour the count rate had fallen to 80 counts min^{-1}.

Calculate the half-life, in minutes, of ^{11}C.

1

(c) ^{11}C is injected into the bloodstream as glucose molecules ($C_6H_{12}O_6$). Some of the carbon atoms in these glucose molecules are ^{11}C atoms.

The intensity of radiation in a sample of ^{11}C is compared with the intensity of radiation in a sample of glucose containing ^{11}C atoms. Both samples have the same mass.

Which sample has the higher intensity of radiation?

Give a reason for your answer.

1

(3)

Marks

7. Hydrogen cyanide, HCN, is highly toxic.

   (a) Information about hydrogen cyanide is given in the table.

Structure	$H-C\equiv N$
Molecular mass	27
Boiling point	26 °C

   Although hydrogen cyanide has a similar molecular mass to nitrogen, it has a much higher boiling point. This is due to the permanent dipole–permanent dipole attractions in liquid hydrogen cyanide.

   What is meant by permanent dipole–permanent dipole attractions?

   Explain how they arise in liquid hydrogen cyanide.

   2

   (b) Hydrogen cyanide is of great importance in organic chemistry. It offers a route to increasing the chain length of a molecule.

   If ethanal is reacted with hydrogen cyanide and the product hydrolysed with acid, lactic acid is formed.

   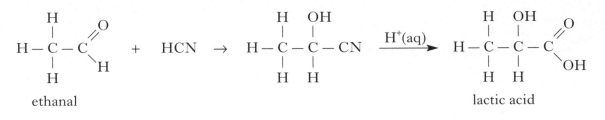

   ethanal   lactic acid

   Draw a structural formula for the acid produced when propanone is used instead of ethanal in the above reaction sequence.

   1

   (3)

Marks

8. Glycerol, $C_3H_8O_3$, is widely used as an ingredient in toothpaste and cosmetics.

(a) Glycerol is mainly manufactured from fats and oils. Propene can be used as a feedstock in an alternative process as shown.

$$H-\underset{\underset{H}{|}}{C}=\underset{\underset{H}{|}}{C}-\underset{\underset{H}{|}}{\overset{\overset{H}{|}}{C}}-H$$

propene

Stage 1 ↓

$$H-\underset{\underset{H}{|}}{C}=\underset{\underset{H}{|}}{C}-\underset{\underset{H}{|}}{\overset{\overset{Cl}{|}}{C}}-H \quad \xrightarrow[\text{ClOH}]{\textbf{Stage 2}} \quad H-\underset{\underset{H}{|}}{\overset{\overset{Cl}{|}}{C}}-\underset{\underset{H}{|}}{\overset{\overset{OH}{|}}{C}}-\underset{\underset{H}{|}}{\overset{\overset{Cl}{|}}{C}}-H \quad \xrightarrow[\text{NaOH}]{\textbf{Stage 3}} \quad H-\underset{\underset{H}{|}}{C}\overset{O}{\overset{/\,\backslash}{-}}\underset{\underset{H}{|}}{C}-\underset{\underset{H}{|}}{\overset{\overset{Cl}{|}}{C}}-H$$

Stage 4 ↓

$$H-\underset{\underset{H}{|}}{\overset{\overset{OH}{|}}{C}}-\underset{\underset{H}{|}}{\overset{\overset{OH}{|}}{C}}-\underset{\underset{H}{|}}{\overset{\overset{OH}{|}}{C}}-H$$

glycerol

(i) What is meant by a feedstock?

1

(ii) Name the type of reaction taking place in **Stage 2**.

1

(iii) In **Stage 3**, a salt and water are produced as by-products.

Name the salt produced.

1

Marks

8. **(a)** **(continued)**

(iv) Apart from cost, state **one** advantage of using fats and oils rather than propene in the manufacture of glycerol.

1

(b) Hydrogen has been named as a 'fuel for the future'. In a recent article researchers reported success in making hydrogen from glycerol:

$$C_3H_8O_3(\ell) \rightarrow CO_2(g) + CH_4(g) + H_2(g)$$

Balance this equation.

1

(c) The enthalpy of formation of glycerol is the enthalpy change for the reaction:

$$3C(s) + 4H_2(g) + 1\frac{1}{2}O_2(g) \rightarrow C_3H_8O_3(\ell)$$
(graphite)

Calculate the enthalpy of formation of glycerol, in kJ mol^{-1}, using information from the data booklet and the following data.

$$C_3H_8O_3(\ell) + 3\frac{1}{2}O_2(g) \rightarrow 3CO_2(g) + 4H_2O(\ell) \quad \Delta H = -1654 \, kJ \, mol^{-1}$$

Show your working clearly.

2

(7)

[Turn over

Marks

9. Enzymes are biological catalysts.

(*a*)   Name the **four** elements present in all enzymes.

1

(*b*)   The enzyme catalase, found in potatoes, can catalyse the decomposition of hydrogen peroxide.

$$2H_2O_2(aq) \rightarrow 2H_2O(\ell) + O_2(g)$$

A student carried out the Prescribed Practical Activity (PPA) to determine the effect of pH on enzyme activity.

Describe how the activity of the enzyme was measured in this PPA.

1

(*c*)   A student wrote the following **incorrect** statement.

When the temperature is increased, enzyme-catalysed reactions will always speed up because more molecules have kinetic energy greater than the activation energy.

Explain the mistake in the student's reasoning.

enzyme catalysed reactions will speed up, however, after they will react to d0 because

1

(3)

Marks

10. Sulphur trioxide can be prepared in the laboratory by the reaction of sulphur dioxide with oxygen.

$$2SO_2(g) + O_2(g) \rightleftharpoons 2SO_3(g)$$

The sulphur dioxide and oxygen gases are dried by bubbling them through concentrated sulphuric acid. The reaction mixture is passed over heated vanadium(V) oxide.

Sulphur trioxide has a melting point of 17 °C. It is collected as a white crystalline solid.

(*a*) Complete the diagram to show how the reactant gases are dried and the product is collected.

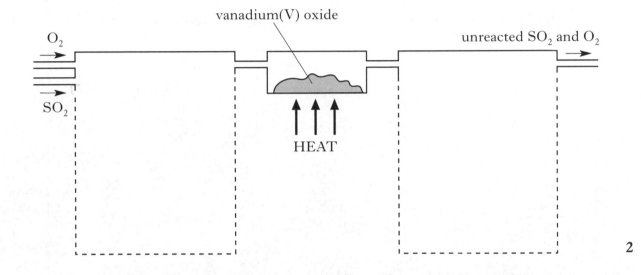

(*b*) Under certain conditions, 43·2 tonnes of sulphur trioxide are produced in the reaction of 51·2 tonnes of sulphur dioxide with excess oxygen.

Calculate the percentage yield of sulphur trioxide.

Show your working clearly.

2

2

(4)

[Turn over

Marks

11. (a) The first ionisation energy of an element is defined as the energy required to remove one mole of electrons from one mole of atoms in the gaseous state.

The graph shows the first ionisation energies of the Group 1 elements.

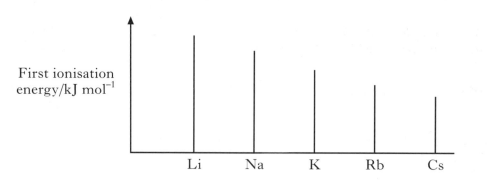

(i) Clearly explain why the first ionisation energy decreases down this group.

2

(ii) The energy needed to remove one electron from one helium atom is 3.94×10^{-21} kJ.

Calculate the first ionisation energy of helium, in kJ mol^{-1}.

1

(b) The ability of an atom to form a negative ion is measured by its Electron Affinity.

The Electron Affinity is defined as the energy change when one mole of gaseous atoms of an element combines with one mole of electrons to form gaseous negative ions.

Write the equation, showing state symbols, that represents the Electron Affinity of chlorine.

1

(4)

Marks

12. (*a*) A student bubbled $240 \, cm^3$ of carbon dioxide into $400 \, cm^3$ of $0 \cdot 10 \, mol \, l^{-1}$ lithium hydroxide solution.

The equation for the reaction is:

$$2LiOH(aq) \; + \; CO_2(g) \; \rightarrow \; Li_2CO_3(aq) \; + \; H_2O(\ell)$$

Calculate the number of moles of lithium hydroxide that would **not** have reacted.

(Take the molar volume of carbon dioxide to be 24 litres mol^{-1}.)

Show your working clearly.

2

(*b*) What is the pH of the $0 \cdot 10 \, mol \, l^{-1}$ lithium hydroxide solution used in the experiment?

1

(*c*) Explain why lithium carbonate solution has a pH greater than 7.

In your answer you should mention the **two** equilibria involved.

2

(5)

DO NOT WRITE IN THIS MARGIN

Marks

13. (*a*) A sample of petrol was analysed to identify the hydrocarbons present. The results are shown in the table.

Number of carbon atoms per molecule	**Hydrocarbons present in the sample**
4	2-methylpropane
5	2-methylbutane
6	2,3-dimethylbutane
7	2,2-dimethylpentane 2,2,3-trimethylbutane

(i) Draw a structural formula for 2,2,3-trimethylbutane. **1**

(ii) The structures of the hydrocarbons in the sample are similar in a number of ways.

What similarity in structure makes these hydrocarbons suitable for use in unleaded petrol?

branched

1

(*b*) In some countries, organic compounds called 'oxygenates' are added to unleaded petrol.

One such compound is MTBE.

MTBE $$\begin{array}{c} \quad\quad CH_3 \\ \quad\quad | \\ H_3C - C - O - CH_3 \\ \quad\quad | \\ \quad\quad CH_3 \end{array}$$

(i) Suggest why oxygenates such as MTBE are added to unleaded petrol. **1**

Marks

13. (*b*) **(continued)**

(ii) MTBE is an example of an ether. All ethers contain the functional group:

$CH_3 OCH_3$

Draw a structural formula for an isomer of MTBE that is also an ether.

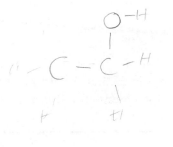

1

(*c*) Some of the hydrocarbons that are suitable for unleaded petrol are produced by a process known as reforming.

One reforming reaction is:

$$H - C - C - C - C - C - C - H \rightarrow \text{hydrocarbon } \mathbf{A} + H_2$$

hexane

Hydrocarbon **A** is non-aromatic and does **not** decolourise bromine solution.

Give a possible name for hydrocarbon **A**.

1

(5)

[Turn over

Marks

14. (*a*) Hess's Law can be verified using the reactions summarised below.

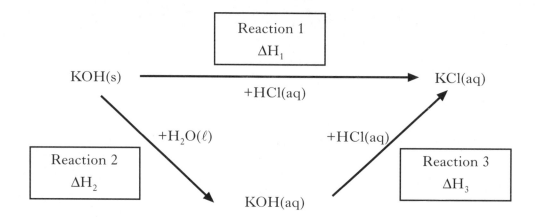

(i) Complete the list of measurements that would have to be carried out in order to determine the enthalpy change for Reaction 2.

Reaction 2

1. Using a measuring cylinder, measure out $25\,cm^3$ of water into a polystyrene cup.

2.

3. Weigh out accurately about $1\cdot2\,g$ of potassium hydroxide and add it to the water, with stirring, until all the solid dissolves.

4.

1

(ii) Why was the reaction carried out in a polystyrene cup?

1

Marks

14. **(a)** **(continued)**

(iii) A student found that $1 \cdot 08 \, kJ$ of energy was **released** when $1 \cdot 2 \, g$ of potassium hydroxide was dissolved completely in water.

Calculate the enthalpy of solution of potassium hydroxide.

1

(b) A student wrote the following **incorrect** statement.

The enthalpy of neutralisation for hydrochloric acid reacting with potassium hydroxide is less than that for sulphuric acid reacting with potassium hydroxide because fewer moles of water are formed as shown in these equations.

$$HCl \; + \; KOH \; \rightarrow \; KCl \; + \; H_2O$$

$$H_2SO_4 \; + \; 2KOH \; \rightarrow \; K_2SO_4 \; + \; 2H_2O$$

Explain the mistake in the student's statement.

1

(4)

[Turn over

Marks

15. Infra-red spectroscopy is a technique that can be used to identify the bonds that are present in a molecule.

Different bonds absorb infra-red radiation of different wavenumbers. This is due to differences in the bond 'stretch'. These absorptions are recorded in a spectrum.

A spectrum for propan-1-ol is shown.

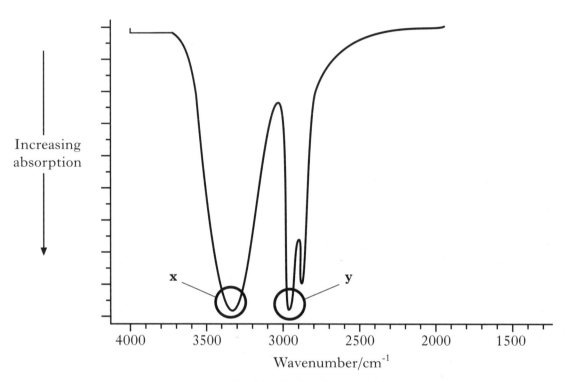

The correlation table on page 13 of the data booklet shows the wavenumber ranges for the absorptions due to different bonds.

(a) Use the correlation table to identify the bonds responsible for the two absorptions, **x** and **y**, that are circled in the propan-1-ol spectrum.

x: **y**:

1

(b) Propan-1-ol reacts with ethanoic acid.

(i) What name is given to this type of reaction?

1

Marks

15. (*b*) **(continued)**

(ii) Draw a spectrum that could be obtained for the organic product of this reaction.

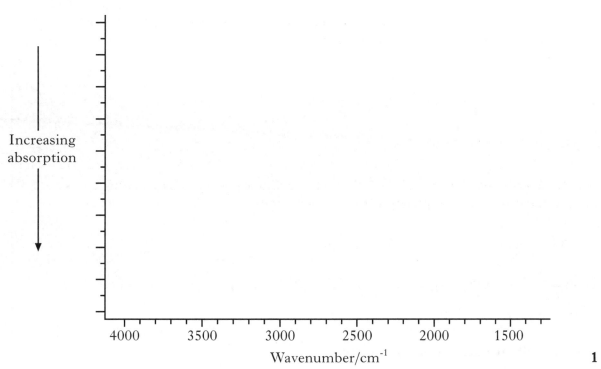

1

(3)

[Turn over

Marks

16. A major problem for the developed world is the pollution of rivers and streams by nitrite and nitrate ions.

   The concentration of nitrite ions, $NO_2^-(aq)$, in water can be determined by titrating samples against acidified permanganate solution.

   (a) Suggest **two** points of good practice that should be followed to ensure that an accurate end-point is achieved in a titration.

   1

   (b) An average of $21 \cdot 6\,cm^3$ of $0 \cdot 0150\,mol\,l^{-1}$ acidified permanganate solution was required to react completely with the nitrite ions in a $25 \cdot 0\,cm^3$ sample of river water.

   The equation for the reaction taking place is:

   $$2MnO_4^-(aq) + 5NO_2^-(aq) + 6H^+(aq) \rightarrow 2Mn^{2+}(aq) + 5NO_3^-(aq) + 3H_2O(\ell)$$

   (i) Calculate the nitrite ion concentration, in $mol\,l^{-1}$, in the river water.

   Show your working clearly.

   2

   (ii) During the reaction the nitrite ion is oxidised to the nitrate ion.

   Complete the ion-electron equation for the oxidation of the nitrite ions.

   $$NO_2^-(aq) \quad \rightarrow \quad NO_3^-(aq)$$

   1

   (4)

   [END OF QUESTION PAPER]

HIGHER

2011

[BLANK PAGE]

FOR OFFICIAL USE

Total
Section B

X012/301

NATIONAL
QUALIFICATIONS
2011

THURSDAY, 26 MAY
9.00 AM – 11.30 AM

CHEMISTRY
HIGHER

Fill in these boxes and read what is printed below.

Full name of centre

Town

Forename(s)

Surname

Date of birth

Day	Month	Year	Scottish candidate number	Number of seat

Reference may be made to the Chemistry Higher and Advanced Higher Data Booklet.

SECTION A—Questions 1–40 (40 marks)

Instructions for completion of **Section A** are given on page two.

For this section of the examination you must use an **HB pencil**.

SECTION B (60 marks)

1   All questions should be attempted.

2   The questions may be answered in any order but all answers are to be written in the spaces provided in this answer book, **and must be written clearly and legibly in ink**.

3   Rough work, if any should be necessary, should be written in this book and then scored through when the fair copy has been written. If further space is required, a supplementary sheet for rough work may be obtained from the Invigilator.

4   Additional space for answers will be found at the end of the book. If further space is required, supplementary sheets may be obtained from the Invigilator and should be inserted inside the **front** cover of this book.

5   The size of the space provided for an answer should not be taken as an indication of how much to write. It is not necessary to use all the space.

6   Before leaving the examination room you must give this book to the Invigilator. If you do not, you may lose all the marks for this paper.

SECTION A

Read carefully

1 Check that the answer sheet provided is for **Chemistry Higher (Section A)**.

2 For this section of the examination you must use an **HB pencil** and, where necessary, an eraser.

3 Check that the answer sheet you have been given has **your name**, **date of birth**, **SCN** (Scottish Candidate Number) and **Centre Name** printed on it.

Do not change any of these details.

4 If any of this information is wrong, tell the Invigilator immediately.

5 If this information is correct, **print** your name and seat number in the boxes provided.

6 The answer to each question is **either** A, B, C or D. Decide what your answer is, then, using your pencil, put a horizontal line in the space provided (see sample question below).

7 There is **only one correct** answer to each question.

8 Any rough working should be done on the question paper or the rough working sheet, **not** on your answer sheet.

9 At the end of the examination, put the **answer sheet for Section A inside the front cover of your answer book**.

Sample Question

To show that the ink in a ball-pen consists of a mixture of dyes, the method of separation would be

        A chromatography

        B fractional distillation

        C fractional crystallisation

        D filtration.

The correct answer is **A**—chromatography. The answer **A** has been clearly marked in **pencil** with a horizontal line (see below).

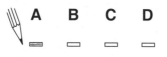

Changing an answer

If you decide to change your answer, carefully erase your first answer and using your pencil, fill in the answer you want. The answer below has been changed to **D**.

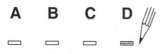

1. Which of the following gases could be described as monatomic?

   A Bromine

   B Methane

   C Hydrogen

   D Helium

2. Different isotopes of the same element have identical

   A electron arrangements

   B nuclei

   C numbers of neutrons

   D mass numbers.

3. Which of the following pairs of solutions will **not** react to produce a precipitate?

   A Copper(II) chloride and lithium sulphate

   B Potassium carbonate and zinc sulphate

   C Silver nitrate and sodium chloride

   D Ammonium phosphate and magnesium chloride

4. Which of the following sugars does **not** react with Benedict's solution?

   A Glucose

   B Fructose

   C Maltose

   D Sucrose

5. Which of the following gases would contain the **greatest** number of molecules in a 100 g sample, at room temperature?

   A Fluorine

   B Hydrogen

   C Nitrogen

   D Oxygen

6.

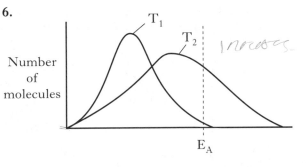

Kinetic energy of molecules

Which line in the table is correct for a reaction as the temperature **decreases** from T_2 to T_1?

	Activation energy (E_A)	Number of successful collisions
A	remains the same	increases
B	decreases	decreases
C	decreases	increases
D	remains the same	decreases

7. A pupil added 0·1 mol of zinc to a solution containing 0·05 mol of silver(I) nitrate.

   $$Zn(s) + 2AgNO_3(aq) \rightarrow Zn(NO_3)_2(aq) + 2Ag(s)$$

   Which of the following statements about the experiment is correct?

   A 0·05 mol of zinc reacts.

   B 0·05 mol of silver is displaced.

   C Silver nitrate is in excess.

   D All of the zinc reacts.

8. A reaction takes place in two stages.

   Stage 1

   $$S_2O_8^{2-}(aq) + 2I^-(aq) + 2Fe^{2+}(aq) \rightarrow 2SO_4^{2-}(aq) + 2I^-(aq) + 2Fe^{3+}(aq)$$

   Stage 2

   $$2SO_4^{2-}(aq) + 2I^-(aq) + 2Fe^{3+}(aq) \rightarrow 2SO_4^{2-}(aq) + I_2(aq) + 2Fe^{2+}(aq)$$

   The ion that is the catalyst in the reaction is

   A $S_2O_8^{2-}(aq)$

   B $I^-(aq)$

   C $Fe^{2+}(aq)$

   D $SO_4^{2-}(aq)$.

9. The following potential diagram is for a reaction carried out with and without a catalyst.

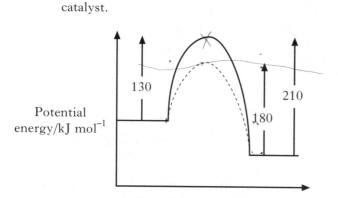

Reaction path

The activation energy for the catalysed reaction is

A   30 kJ mol^{-1}

B   80 kJ mol^{-1}

C   100 kJ mol^{-1}

D   130 kJ mol^{-1}.

10. Which of the following equations represents an enthalpy of combustion?

A $C_2H_6(g) + 3\frac{1}{2}O_2(g)$
    ↓
    $2CO_2(g) + 3H_2O(\ell)$

B $C_2H_5OH(\ell) + O_2(g)$
    ↓
    $CH_3COOH(\ell) + H_2O(\ell)$

C $CH_3CHO(\ell) + \frac{1}{2}O_2(g)$
    ↓
    $CH_3COOH(\ell)$

D $CH_4(g) + 1\frac{1}{2}O_2(g)$
    ↓
    $CO(g) + 2H_2O(\ell)$

11. A potassium atom is larger than a sodium atom because potassium has

A   a larger nuclear charge

B   a larger nucleus

C   more occupied energy levels

D   a smaller ionisation energy.

12. Hydrogen will form a non-polar covalent bond with an element which has an electronegativity value of

A   0·9

B   1·5

C   2·2

D   2·5.

13. Which property of a chloride would prove that it contained ionic bonding?

A   It conducts electricity when molten.

B   It is soluble in a polar solvent.

C   It is a solid at room temperature.

D   It has a high boiling point.

14. A mixture of potassium chloride and potassium carbonate is known to contain 0·1 mol of chloride ions and 0·1 mol of carbonate ions.

How many moles of potassium ions are present?

A   0·15

B   0·20

C   0·25

D   0·30

15. Which of the following has the largest volume under the same conditions of temperature and pressure?

A   1 g hydrogen

B   14 g nitrogen

C   20·2 g neon

D   35·5 g chlorine

16. $20\,cm^3$ of butane is burned in $150\,cm^3$ of oxygen.

$$C_4H_{10}(g) + 6\frac{1}{2}O_2(g) \rightarrow 4CO_2(g) + 5H_2O(g)$$

What is the total volume of gas present after complete combustion of the butane?

A $80\,cm^3$

B $100\,cm^3$

C $180\,cm^3$

D $200\,cm^3$

17. Which of the following types of hydrocarbons when added to petrol would **not** reduce "knocking"?

   A Cycloalkanes

   B Aromatic hydrocarbons

   C Branched-chain alkanes

   D Straight-chain alkanes

18. Which of the following fuels when burned would make no contribution to global warming?

   A Hydrogen

   B Natural gas

   C Petrol

   D Coal

19. Which of the following hydrocarbons always gives the same product when one of its hydrogen atoms is replaced by a chlorine atom?

   A Hexane

   B Hex-1-ene

   C Cyclohexane

   D Cyclohexene

20.

The name of this compound is

   A methanol

   B methanal

   C methanoic acid

   D methanone.

21. Which of the following is an isomer of ethyl propanoate?

   A Pentan-2-one

   B Pentanoic acid

   C Methyl propanoate

   D Pentane-1,2-diol

22. Propan-2-ol can be prepared from propane as follows.

$$CH_3 - CH_2 - CH_3 \xrightarrow{\text{Step } 1} CH_3 - CH = CH_2 \xrightarrow{\text{Step } 2} CH_3 - \underset{\underset{OH}{|}}{CH} - CH_3$$

Which line in the table correctly describes the types of reaction taking place at Steps **1** and **2**?

	Step 1	Step 2
A	cracking	hydration
B	cracking	hydrolysis
C	dehydration	hydration
D	dehydration	hydrolysis

[Turn over

23.

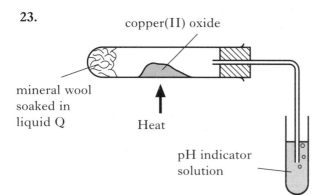

After heating for several minutes, as shown in the diagram, the pH indicator solution turned red.

Liquid **Q** could be

A   propanone

B   paraffin

C   propan-1-ol

D   propan-2-ol.

24. Which of the following compounds is hydrolysed when warmed with sodium hydroxide solution?

$$A \quad CH_3 - \overset{\overset{\displaystyle O}{\|}}{C} - CH_2 - CH_3$$

$$B \quad CH_3 - CH_2 - \overset{\overset{\displaystyle O}{\|}}{C} - O - CH_3$$

$$C \quad CH_3 - \overset{\overset{\displaystyle OH}{|}}{CH} - CH_2 - CH_3$$

$$D \quad CH_3 - CH_2 - \underset{\underset{\displaystyle CH_3}{|}}{CH} - \overset{\overset{\displaystyle O}{\|}}{C} - H$$

25. Which of the following is most likely to be used as a flavouring?

A CH_3CH_2CHO

B $CH_3CH_2CH_2COOH$

C $CH_3CH(OH)CH_2CH_3$

D $CH_3CH_2CH_2COOCH_2CH_3$

26. Thermosetting plastics can be made by the following sequence of reactions.

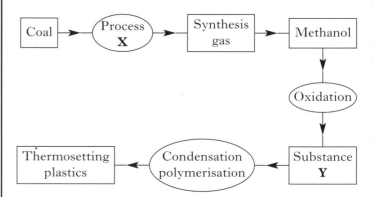

Which line in the table shows the correct names for Process **X** and Substance **Y**?

	Process X	Substance Y
A	combustion	methanal
B	combustion	methanoic acid
C	steam reforming	methanoic acid
D	steam reforming	methanal

27. Which of the following is an amine?

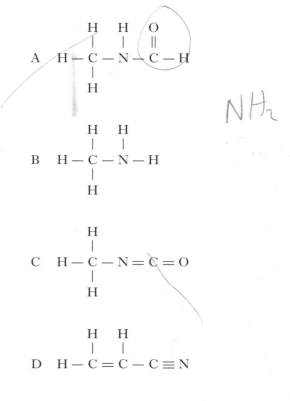

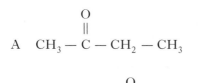

28. Noradrenaline and phenylephrine cause increases in the blood pressure because the part of each of these molecules that they have in common has the correct shape to allow them to bind to a certain human protein.

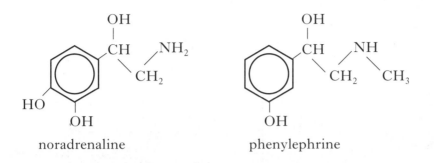

noradrenaline   phenylephrine

The part of these molecules which is the correct shape to bind to the protein is

A 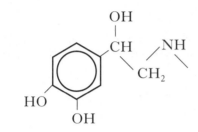   B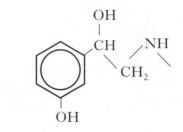

C 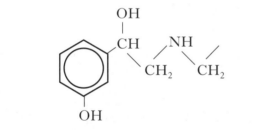   D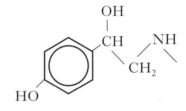

---

29. Which of the following polymers is photoconductive?

  A Kevlar

  B Biopol

  C Poly(ethenol)

  D Poly(vinyl carbazole)

30. Which of the following compounds can be classified as proteins?

  A Fats

  B Oils

  C Enzymes

  D Amino acids

31. The flow chart summarises some industrial processes involving ethene.

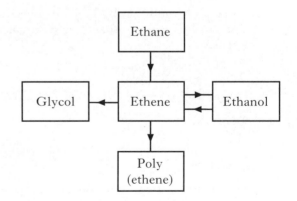

The feedstocks for ethene in these processes are

  A ethane and glycol

  B ethane and ethanol

  C glycol and poly(ethene)

  D glycol, poly(ethene) and ethanol.

32. The enthalpy change for $K(s) \rightarrow K(g)$ is $88\,kJ\,mol^{-1}$.

Using the above information and information from the data booklet (page 10), the enthalpy change for $K(s) \rightarrow K^{2+}(g) + 2e^-$ is

A $513\,kJ\,mol^{-1}$

B $3060\,kJ\,mol^{-1}$

C $3485\,kJ\,mol^{-1}$

D $3573\,kJ\,mol^{-1}$.

33. Two flasks, A and B, were placed in a water bath at $40\,^\circ C$.

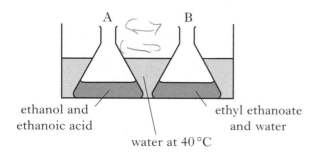

ethanol and ethanoic acid

ethyl ethanoate and water

water at $40\,^\circ C$

After several days the contents of both flasks were analysed.

Which result would be expected?

A   Flask A contains ethyl ethanoate, water, ethanol and ethanoic acid; flask B is unchanged.

B   Flask A contains only ethyl ethanoate and water; flask B is unchanged.

C   Flask A contains only ethyl ethanoate and water; flask B contains ethyl ethanoate, water, ethanol and ethanoic acid.

D   Flask A and flask B contain ethyl ethanoate, water, ethanol and ethanoic acid.

34. $NH_3(g) + H_2O(\ell) \rightleftharpoons NH_4^+(aq) + OH^-(aq)$

$\Delta H = -36\,kJ\,mol^{-1}$

The solubility of ammonia in water will be increased by

A   increasing pressure and cooling

B   decreasing pressure and cooling

C   decreasing pressure and warming

D   increasing pressure and warming.

35. The pH of a $0{\cdot}1\,mol\,l^{-1}$ solution of an acid was measured and found to be pH 4.

The pH of a $0{\cdot}001\,mol\,l^{-1}$ solution of an alkali was measured and found to be pH 11.

Which line in the table is correct?

	Acid	Alkali
A	weak	weak
B	weak	strong
C	strong	weak
D	strong	strong

36. Which of the following solutions contains equal concentrations of $H^+(aq)$ and $OH^-(aq)$ ions?

A $NH_4Cl(aq)$

B $Na_2CO_3(aq)$

C $KNO_3(aq)$

D $CH_3COOK(aq)$

37. During a redox process in acid solution, iodate ions are converted into iodine.

$2IO_3^-(aq) + 12H^+(aq) + xe^- \rightarrow I_2(aq) + 6H_2O(\ell)$

To balance the equation, what is the value of **x**?

A   2

B   6

C   10

D   12

38. The following reactions take place when nitric acid is added to zinc.

$NO_3^-(aq) + 4H^+(aq) + 3e^- \rightarrow NO(g) + 2H_2O(\ell)$

$Zn(s) \rightarrow Zn^{2+}(aq) + 2e^-$

How many moles of $NO_3^-(aq)$ are reduced by one mole of zinc?

A $\dfrac{2}{3}$

B   1

C $\dfrac{3}{2}$

D   2

39. If 96 500 C of electricity are passed through separate solutions of copper(II) chloride and nickel(II) chloride, then

A equal masses of copper and nickel will be deposited

B the same number of atoms of each metal will be deposited

C the metals will be plated on the positive electrode

D different numbers of moles of each metal will be deposited.

40. Some smoke detectors make use of radiation which is very easily stopped by tiny smoke particles moving between the radioactive source and the detector.

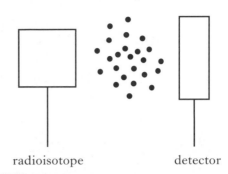

radioisotope   detector

The most suitable type of radioisotope for a smoke detector would be

A an alpha-emitter with a long half-life

B a gamma-emitter with a short half-life

C an alpha-emitter with a short half-life

D a gamma-emitter with a long half-life.

Candidates are reminded that the answer sheet MUST be returned INSIDE the front cover of this answer book.

[Turn over

SECTION B

All answers must be written clearly and legibly in ink.

1. Chloromethane, CH_3Cl, can be produced by reacting methanol solution with dilute hydrochloric acid using a solution of zinc chloride as a catalyst.

$$CH_3OH(aq) + HCl(aq) \xrightarrow{ZnCl_2(aq)} CH_3Cl(aq) + H_2O(\ell)$$

(a) What type of catalysis is taking place?

homogeneous

1

(b) The graph shows how the concentration of the hydrochloric acid changed over a period of time when the reaction was carried out at 20 °C.

Concentration of acid/mol l^{-1} (y-axis, values 0·00 to 1·80)

Time/min (x-axis, values 0 to 2000)

(i) Calculate the average rate, in mol l^{-1} min^{-1}, in the first 400 minutes.

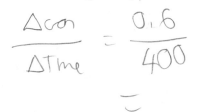

$$\frac{\Delta con}{\Delta Time} = \frac{0.6}{400}$$

$$=$$

1

(ii) On the graph above, sketch a curve to show how the concentration of hydrochloric acid would change over time if the reaction is repeated at 30 °C.

(Additional graph paper, if required, can be found on *Page thirty-five*).

1

(3)

DO NOT WRITE IN THIS MARGIN

Marks

2. The elements from sodium to argon make up the third period of the Periodic Table.

(*a*) On crossing the third period from left to right there is a general increase in the first ionisation energy of the elements.

(i) Why does the first ionisation energy increase across the period? **1**

(ii) Write an equation corresponding to the first ionisation energy of chlorine. **1**

(*b*) The electronegativities of elements in the third period are listed on page 10 of the databook.

Why is no value provided for the noble gas, argon?

It is so unreactive it does not react with anything

1

(3)

[Turn over

Marks

3. A student writes the following two statements. **Both are incorrect**. In each case explain the mistake in the student's reasoning.

(a) All ionic compounds are solids at room temperature. Many covalent compounds are gases at room temperature. This proves that ionic bonds are stronger than covalent bonds.

1

(b) The formula for magnesium chloride is $MgCl_2$ because, in solid magnesium chloride, each magnesium ion is bonded to two chloride ions.

1

(2)

Marks

4. Petrol is a complex blend of many chemicals.

(a) A typical hydrocarbon found in petrol is shown below.

What is the systematic name for this compound?

2,2,4-trimethylpentane

1

(b) In what way is a petrol that has been blended for use in winter different from a summer blend?

— Winter, small more volatile hydrocarbons

— summer, bigger less volatile hydrocarbons

1

(c) The ester methyl stearate is also a useful vehicle fuel.

A student prepared this ester from methanol and stearic acid during the Prescribed Practical Activity, "Making Esters".

Describe how this ester was prepared.

2

(4)

Marks

5. Chlorine gas can be produced by heating calcium hypochlorite, $Ca(OCl)_2$, in dilute hydrochloric acid.

$$Ca(OCl)_2(s) + 2HCl(aq) \rightarrow Ca(OH)_2(aq) + 2Cl_2(g)$$

(a) Calculate the mass of calcium hypochlorite that would be needed to produce 0·096 litres of chlorine gas.

(Take the molar volume of chlorine gas to be 24 litres mol^{-1}.)

Show your working clearly.

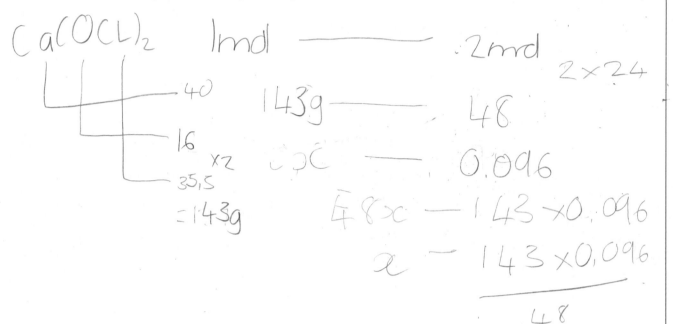

2

Marks

5. **(continued)**

(b) Chlorine is used in the manufacture of herbicides such as
2,4-dichlorophenoxyethanoic acid.

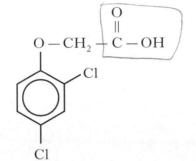

Another commonly used herbicide is 4-chloro-2-methylphenoxyethanoic acid.

Draw a structural formula for 4-chloro-2-methylphenoxyethanoic acid.

1

(3)

[Turn over

Marks

6.   Hairspray is a mixture of chemicals.

(*a*)   A primary alcohol, 2-methylpropan-1-ol, is added to hairspray to help it dry quickly on the hair.

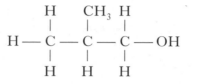

Draw a structural formula for a secondary alcohol that is an isomer of 2-methylpropan-1-ol.

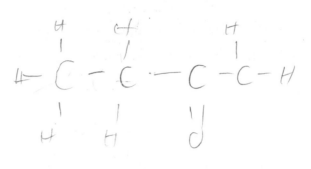

1

Marks

(b) Triethanol amine and triisopropyl amine are bases used to neutralise acidic compounds in the hairspray to prevent damage to the hair.

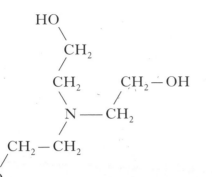

triethanol amine	triisopropyl amine
molecular mass 149	molecular mass 143
boiling point 335 °C	boiling point 47 °C

In terms of the intermolecular bonding present, **explain clearly** why triethanol amine has a much higher boiling point than triisopropyl amine.

OH → hydrogen bonding which is extremely strong here more energy is needed to break

due to tri being bigger van der waals are present more and hence increase bp.

2

(3)

[Turn over

Marks

7. Paracetamol is a widely used painkiller.

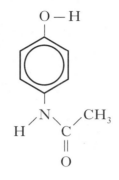

(*a*) Write the molecular formula for paracetamol.

$C_8H_{11}NO_2$

1

(*b*) One antidote for paracetamol overdose is methionine.

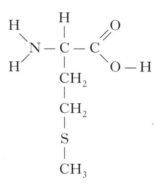

To what family of organic compounds does methionine belong?

amino acids

1

Marks

7.   (continued)

(c)   The concentration of paracetamol in a solution can be determined by measuring how much UV radiation it absorbs.

The graph shows how the absorbance of a sample containing $0.040\,g\,l^{-1}$ paracetamol varies with wavelength.

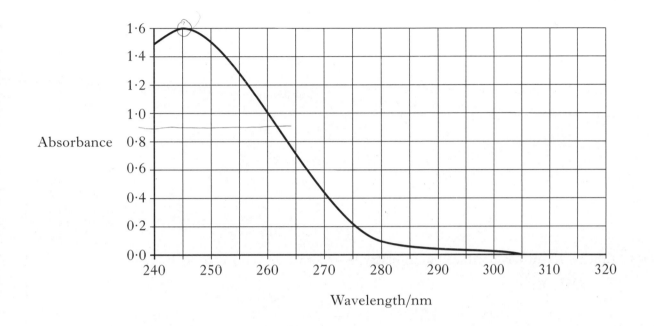

Wavelength/nm

The quantity of UV radiation of wavelength 245 nm absorbed is directly proportional to the concentration of paracetamol.

The absorbance of a second sample of paracetamol solution measured at 245 nm was 0·90.

What is the concentration, in $g\,l^{-1}$, of this second paracetamol solution?

$$1.6 = 245 = 0.040$$

$$0.9 - 245 -$$

1

(3)

[Turn over

Marks

8. Diols are widely used in the manufacture of polyester polymers.

Polyethylene naphthalate is used to manufacture food containers. The monomers used to produce this polymer are shown.

$$O \diagdown C - C_{10}H_6 - C \diagup O$$
$$HO \diagup \qquad\qquad \diagdown OH$$

naphthalenedicarboxylic acid

$$HO - CH_2 - CH_2 - OH$$

ethane-1,2-diol

(a) Draw the repeating unit for polyethylene naphthalate.

1

(b) Ethane-1,2-diol is produced in industry by reacting glycerol with hydrogen.

$$H - \underset{\underset{H}{|}}{\overset{\overset{OH}{|}}{C}} - \underset{\underset{H}{|}}{\overset{\overset{OH}{|}}{C}} - \underset{\underset{H}{|}}{\overset{\overset{OH}{|}}{C}} - H \;+\; H_2 \;\rightarrow\; HO - \underset{\underset{H}{|}}{\overset{\overset{H}{|}}{C}} - \underset{\underset{H}{|}}{\overset{\overset{H}{|}}{C}} - OH \;+\; CH_3OH$$

glycerol   ethane-1,2-diol

Excess hydrogen reacts with 27·6 kg of glycerol to produce 13·4 kg of ethane-1,2-diol.

Calculate the percentage yield of ethane-1,2-diol.

Show your working clearly.

1 mol —— 1 mol

91g —— 62g

27·6 kg ×—— x

91x = 1711.2

x = 18.804

$$\% = \frac{13.4}{18.8} \times 100$$

= 71.25

2

(3)

Marks

9. When vegetable oils are hydrolysed, mixtures of fatty acids are obtained. The fatty acids can be classified by their degree of unsaturation.

   The table below shows the composition of each of the mixtures of fatty acids obtained when palm oil and olive oil were hydrolysed.

	Palm oil	Olive oil
Saturated fatty acids	51%	16%
Monounsaturated fatty acids	39%	75%
Polyunsaturated fatty acids	10%	9%

   (a) Why does palm oil have a higher melting point than olive oil?

   It has more saturated bonds than oils, meaning more energy is needed to break up them

   1

   (b) One of the fatty acids produced by the hydrolysis of palm oil is linoleic acid, $C_{17}H_{31}COOH$.

   To which class (saturated, monounsaturated or polyunsaturated) does this fatty acid belong?

   1

   (c) When a mixture of palm oil and olive oil is hydrolysed using a solution of sodium hydroxide, a mixture of the sodium salts of the fatty acids is obtained.

   State a use for these fatty acid salts.

   Soap

   1

   (3)

[Turn over

Marks

10. Christian Schoenbein discovered ozone, O_3, in 1839.

(*a*) Ozone in air can be detected using paper strips that have been soaked in a mixture of starch and potassium iodide solution. The paper changes colour when ozone is present.

Ozone reacts with potassium iodide and water to form iodine, oxygen and potassium hydroxide.

  (i) Write the balanced chemical equation for this reaction.

1

  (ii) What colour would be seen on the paper when ozone is present?

1

(*b*) Ozone and oxygen gases are produced at the same electrode during the electrolysis of dilute sulphuric acid.

The ion-electron equation for the production of ozone is:

$$3H_2O(\ell) \rightarrow O_3(g) + 6H^+(aq) + 6e^-$$

Draw a labelled diagram of the assembled apparatus that could be used to carry out the electrolysis of dilute sulphuric acid, showing how the ozone/oxygen gas mixture can be collected.

2

DO NOT
WRITE IN
THIS
MARGIN

Marks

10. **(continued)**

(*c*) When ozone is bubbled through a solution containing an alkene, an ozonolysis reaction takes place.

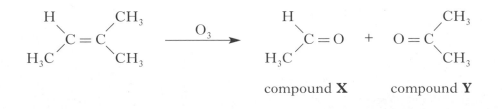

compound **X**   compound **Y**

(i) $2\,cm^3$ of an oxidising agent was added to $5\,cm^3$ of compound **X** in a test tube. After a few minutes a colour change from orange to green was observed.

Name the oxidising agent used.

1

(ii) Draw a structural formula for the alkene which, on ozonolysis, would produce propanal and butan-2-one.

1

(6)

DO NOT WRITE IN THIS MARGIN

Marks

11. Sulphurous acid, H_2SO_3, is a weak acid produced when sulphur-containing compounds in fuels are burned.

(*a*) What is meant by a **weak** acid?

only partially dissociates
not full

1

(*b*) The table below shows the results of two experiments that were carried out to compare sulphurous acid with the strong acid, hydrochloric acid.

	Sulphurous acid 0·1 mol l^{-1}	**Hydrochloric acid 0·1 mol l^{-1}**
Rate of reaction with strip of magnesium	slow	fast
Volume of acid required to neutralise 20 cm^3 of 0·1 mol l^{-1} sodium hydroxide solution	10 cm^3	20 cm^3

(i) Why did the magnesium react more quickly with the hydrochloric acid?

More H^+ present

1

(ii) Why is a smaller volume of sulphurous acid solution needed to neutralise the sodium hydroxide solution?

It has more OH^+ ions already
in it, therefore does not
as much to get to [illegible]

1

DO NOT WRITE IN THIS MARGIN

Marks

11. (*b*) (continued)

(iii) The concentration of the sodium hydroxide solution used was $0{\cdot}1\ \text{mol}\,\text{l}^{-1}$.

Calculate the pH of this solution.

13

1

(4)

[Turn over

Marks

12. The element iodine has only one isotope that is stable. Several of the radioactive isotopes of iodine have medical uses. Iodine-131, for example, is used in the study of the thyroid gland and it decays by beta emission.

   (a) Why are some atoms unstable?

   neutron to proton is not
   the same

   1

   (b) Complete the balanced nuclear equation for the beta decay of iodine-131.

   $$^{131}_{53}\text{I} \rightarrow ^{131}_{54}\text{Xe} \quad ^{0}_{-1}\text{e}$$

   1

   (c) The graph shows how the mass of iodine-131 in a sample changes over a period of time.

   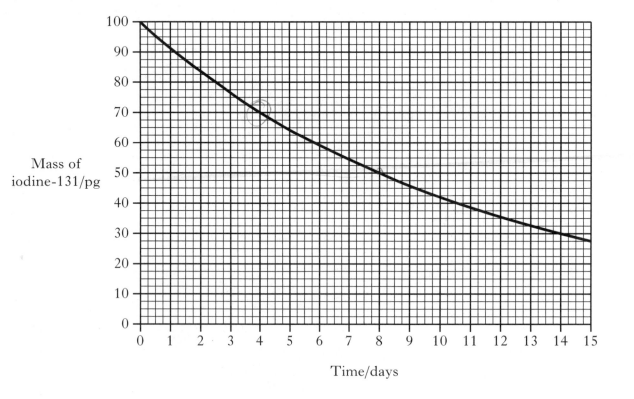

   Mass of iodine-131/pg (y-axis)

   Time/days (x-axis)

   (i) What is the half-life of this isotope?

   8

   1

DO NOT
WRITE IN
THIS
MARGIN

Marks

12. (*c*) **(continued)**

(ii) A sample of sodium iodide solution contained 100 pg of iodine-131 when it was prepared.

Four days later it was injected into a patient.

How many $^{131}I^-$ ions would the 4 day old sample contain?

$(1\,pg = 1 \times 10^{-12}g)$

After 4 days = 70 pg left

$n = \dfrac{m}{gfm}$ $\longrightarrow$ $1 \times 10^{-12} \times 70$

$\longrightarrow$ 126.9

$= \dfrac{7 \times 10^{-11}}{126.9}$

$= 5.52$ mdes

$\begin{array}{cc} 1\,md & 6.02 \times 10^{23} \\ 5.52 & \times \quad x \end{array}$

$= 3.3207$

$= 3.32$ ions

2

(5)

[Turn over

Marks

13. Rivers and drains are carefully monitored to ensure that they remain uncontaminated by potentially harmful substances from nearby industries. Chromate ions, CrO_4^{2-}, are particularly hazardous.

   (a) When chromate ions dissolve in water the following equilibrium is established.

$$2CrO_4^{2-}(aq) \ + \ 2H^+(aq) \ \rightleftharpoons \ Cr_2O_7^{2-}(aq) \ + \ H_2O(\ell)$$
$$\text{yellow} \qquad\qquad\qquad\qquad \text{orange}$$

   Explain fully the colour change that would be observed when solid sodium hydroxide is added to the solution.

2

   (b) The concentration of chromate ions in water can be measured by titrating with a solution of iron(II) sulphate solution.

   (i) To prepare the iron(II) sulphate solution used in this titration, iron(II) sulphate crystals were weighed accurately into a dry beaker.

   Describe how these crystals should be dissolved and then transferred to a standard flask in order to produce a solution of accurately known concentration.

2

Marks

13. (*b*) **(continued)**

(ii) A $50 \cdot 0 \, cm^3$ sample of contaminated water containing chromate ions was titrated and found to require $27 \cdot 4 \, cm^3$ of $0 \cdot 0200 \, mol \, l^{-1}$ iron(II) sulphate solution to reach the end-point.

The redox equation for the reaction is:

$$3Fe^{2+}(aq) \ + \ CrO_4^{2-}(aq) \ + \ 8H^+(aq) \ \rightarrow \ 3Fe^{3+}(aq) \ + \ Cr^{3+}(aq) \ + \ 4H_2O(\ell)$$

Calculate the chromate ion concentration, in $mol \, l^{-1}$, present in the sample of water.

Show your working clearly.

2

(6)

[Turn over

Marks

14. The enthalpies of combustion of some alcohols are shown in the table.

Name of alcohol	Enthalpy of combustion/kJ mol^{-1}
methanol	−727
ethanol	−1367
propan-1-ol	−2020

(a) Using this data, predict the enthalpy of combustion of butan-1-ol, in kJ mol^{-1}.

2686

655
640

1

(b) A value for the enthalpy of combustion of butan-2-ol, C$_4$H$_9$OH, can be determined experimentally using the apparatus shown.

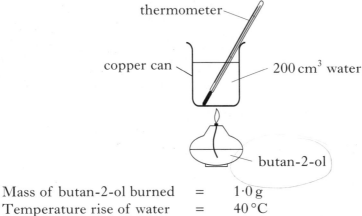

thermometer

copper can

200 cm^3 water

butan-2-ol

Mass of butan-2-ol burned   =   1·0 g
Temperature rise of water   =   40 °C

Use these results to calculate the enthalpy of combustion of butan-2-ol, in kJ mol^{-1}.

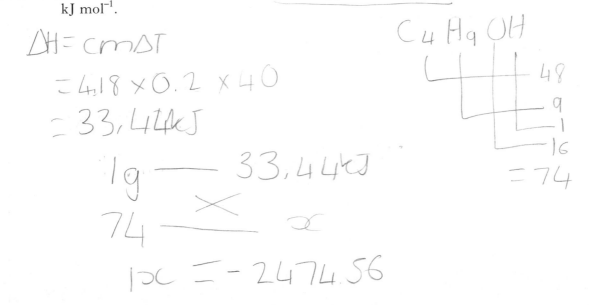

$\Delta H = cm\Delta T$

$= 4.18 \times 0.2 \times 40$

$= 33,44 kJ$

1 g ——— 33,44 kJ

74 ——— x

x = −2474.56

C$_4$H$_9$OH

48
9
1
16
= 74

2

Marks

14. **(continued)**

(*c*) Enthalpy changes can also be calculated using Hess's Law.

The enthalpy of formation for pentan-1-ol is shown below.

$$5C(s) + 6H_2(g) + \frac{1}{2}O_2(g) \rightarrow C_5H_{11}OH(\ell) \qquad \Delta H = -354 \, kJ \, mol^{-1}$$

Using this value, and the enthalpies of combustion of carbon and hydrogen from the data booklet, calculate the enthalpy of combustion of pentan-1-ol, in $kJ \, mol^{-1}$.

$① C(s) + O_2 \longrightarrow CO_2(g) \; \Delta H = -394 \, kJmol^{-1}$

$② H_2(g) + O_2 \longrightarrow H_2O(l) \; \Delta H = -286 \, kJmol$

$③ C_5H_{11}OH + 5O_2 \longrightarrow 2\frac{1}{2}CO_2 + 6H_2O \; \Delta H = -354 \, kJmol$

$5 \times ① \qquad 6 \times ② \qquad -1 \times ③$

$-394 \times 5 \qquad -286 \times 6 \qquad -1 \times -354$

$-1970 \; + \; -1716 \; + \; 354$

$= -3332$

2
(5)

[Turn over

15. Cerium metal is extracted from the mineral monazite.

The flow diagram for the extraction of cerium from the mineral is shown below.

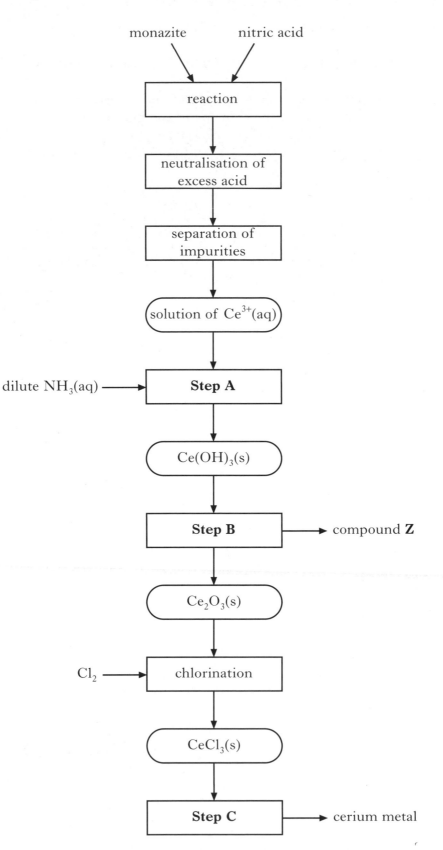

Marks

15. (continued)

(a) Name the type of chemical reaction taking place in **Step A**.

1

(b) In **Step B**, cerium hydroxide is heated to form cerium oxide, Ce_2O_3, and compound **Z**.

Name compound **Z**.

1

(c) In **Step C**, cerium metal is obtained by electrolysis.

(i) What feature of the electrolysis can be used to reduce the cost of cerium production?

1

(ii) The equation for the reaction at the negative electrode is

$$Ce^{3+} + 3e^- \rightarrow Ce$$

Calculate the mass of cerium, in kg, produced in 10 minutes when a current of 4000 A is used.

Show your working clearly.

$$Ce^{3+} + 3e^- \rightarrow Ce$$

$Q = It$

$3 \text{ mol} \quad — \quad 1 \text{ mol}$

$Q = ?$

$289500 \quad — \quad 140.1$

$I = 4000A$

$2400000 \quad — \quad x$

$t = 10 \times 60 = 600s$

$336240000 \quad — \quad 289500 x$

$= 600 \times 4000$

$x = 1161.45 g$

$= 240\ 0000\ C$

2

(5)

[Turn over for Question 16 on *Page thirty-four*

DO NOT WRITE IN THIS MARGIN

Marks

16. The boiling point of water can be raised by the addition of a solute.

The increase in boiling point depends only on the **number** of solute particles but not the type of particle.

The increase in boiling point (ΔT_b), in °C, can be estimated using the formula shown.

$$\Delta T_b = 0{\cdot}51 \times c \times i$$

where

c is the concentration of the solution in mol l^{-1}.
i is the number of particles released into solution when one formula unit of the solute dissolves.

The value of i for a number of compounds is shown in the table below.

Solute	**i**
NaCl	2
$MgCl_2$	3
$(NH_4)_3PO_4$	4

(*a*) What is the value of i for sodium sulphate? **1**

(*b*) Calculate the increase in boiling point, ΔT_b, for a 0·10 mol l^{-1} solution of ammonium phosphate. **1**

(2)

[*END OF QUESTION PAPER*]

ADDITIONAL GRAPH PAPER FOR USE IN QUESTION 1(*b*) (ii)

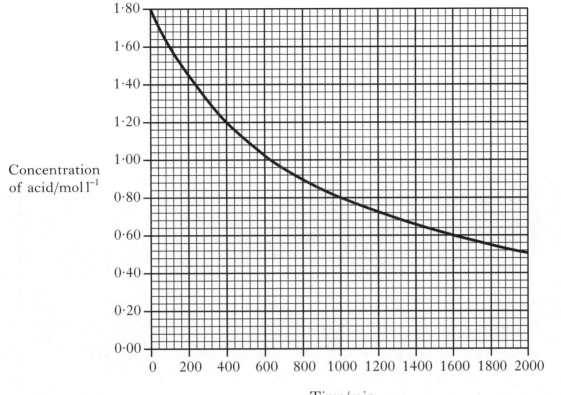

ADDITIONAL SPACE FOR ANSWERS

HIGHER

2012

[BLANK PAGE]

FOR OFFICIAL USE

Total
Section B

X012/12/02

NATIONAL
QUALIFICATIONS
2012

MONDAY, 14 MAY
1.00 PM – 3.30 PM

CHEMISTRY
HIGHER

Fill in these boxes and read what is printed below.

Full name of centre

Town

Forename(s)

Surname

Date of birth

Day	Month	Year	Scottish candidate number	Number of seat

Reference may be made to the Chemistry Higher and Advanced Higher Data Booklet.

SECTION A—Questions 1–40 (40 marks)

Instructions for completion of **Section A** are given on page two.

For this section of the examination you must use an **HB pencil**.

SECTION B (60 marks)

1 All questions should be attempted.

2 The questions may be answered in any order but all answers are to be written in the spaces provided in this answer book, **and must be written clearly and legibly in ink**.

3 Rough work, if any should be necessary, should be written in this book and then scored through when the fair copy has been written. If further space is required, a supplementary sheet for rough work may be obtained from the Invigilator.

4 Additional space for answers will be found at the end of the book. If further space is required, supplementary sheets may be obtained from the Invigilator and should be inserted inside the **front cover** of this book.

5 The size of the space provided for an answer should not be taken as an indication of how much to write. It is not necessary to use all the space.

6 Before leaving the examination room you must give this book to the Invigilator. If you do not, you may lose all the marks for this paper.

SECTION A

Read carefully

1 Check that the answer sheet provided is for **Chemistry Higher (Section A)**.

2 For this section of the examination you must use an **HB pencil** and, where necessary, an eraser.

3 Check that the answer sheet you have been given has **your name**, **date of birth**, **SCN** (Scottish Candidate Number) and **Centre Name** printed on it.

 Do not change any of these details.

4 If any of this information is wrong, tell the Invigilator immediately.

5 If this information is correct, **print** your name and seat number in the boxes provided.

6 The answer to each question is **either** A, B, C or D. Decide what your answer is, then, using your pencil, put a horizontal line in the space provided (see sample question below).

7 There is only **one correct answer** to each question.

8 Any rough working should be done on the question paper or the rough working sheet, **not** on your answer sheet.

9 At the end of the examination, put the **answer sheet for Section A inside the front cover of your answer book**.

Sample Question

To show that the ink in a ball-pen consists of a mixture of dyes, the method of separation would be

 A chromatography

 B fractional distillation

 C fractional crystallisation

 D filtration.

The correct answer is **A**—chromatography. The answer **A** has been clearly marked in **pencil** with a horizontal line (see below).

Changing an answer

If you decide to change your answer, carefully erase your first answer and using your pencil, fill in the answer you want. The answer below has been changed to **D**.

1. Isotopes of an element have

   A the same mass number

   B the same number of neutrons

   C equal numbers of protons and neutrons

   D different numbers of neutrons.

2. Four metals **W**, **X**, **Y** and **Z** and their compounds behaved as described.

   (i) Only **X**, **Y** and **Z** reacted with dilute hydrochloric acid.

   (ii) The oxides of **W**, **X** and **Y** were reduced to the metal when heated with carbon powder. The oxide of **Z** did not react.

   (iii) A displacement reaction occurred when **X** was added to an aqueous solution of the nitrate of **Y**.

   What is the correct order of reactivity of these metals (most reactive first)?

   A W, Y, X, Z

   B W, X, Y, Z

   C Z, X, Y, W

   D Z, Y, X, W

3. A positively charged particle with electron arrangement 2, 8 could be

   A a neon atom

   B a fluoride ion

   C a sodium atom

   D an aluminium ion.

4. A solution of potassium carbonate, made up using tap water, was found to be cloudy.

   This could result from the tap water containing

   A sodium ions

   B chloride ions

   C magnesium ions

   D sulphate ions.

   (You may wish to refer to the Data Booklet.)

5. 1 mol of hydrogen gas and 1 mol of iodine vapour were mixed and allowed to react. After t seconds, 0·8 mol of hydrogen remained.

   The number of moles of hydrogen iodide formed at t seconds was

   A 0·2

   B 0·4

   C 0·8

   D 1·6.

6. In a reaction involving gases, an increase in temperature results in

   A an increase in activation energy

   B an increase in the enthalpy change

   C a decrease in the activation energy

   D more molecules per second forming an activated complex.

7. Calcium carbonate reacts with nitric acid as follows.

   $$CaCO_3(s) + 2HNO_3(aq) \rightarrow Ca(NO_3)_2(aq) + H_2O(\ell) + CO_2(g)$$

   0·05 mol of calcium carbonate was added to a solution containing 0·08 mol of nitric acid.

   Which of the following statements is true?

   A 0·05 mol of carbon dioxide is produced.

   B 0·08 mol of calcium nitrate is produced.

   C Calcium carbonate is in excess by 0·01 mol.

   D Nitric acid is in excess by 0·03 mol.

[Turn over

8.

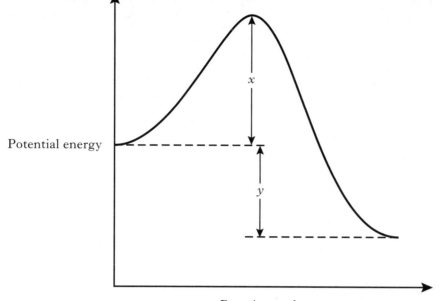

The enthalpy change for the forward reaction can be represented by

A x

B y

C $x + y$

D $x - y$.

9. $5N_2O_4(\ell) + 4CH_3NHNH_2(\ell) \rightarrow 4CO_2(g) + 12H_2O(\ell) + 9N_2(g)$ $\Delta H = -5116\,kJ$

The energy released when 2 moles of each reactant are mixed and ignited is

A $2046\,kJ$

B $2558\,kJ$

C $4093\,kJ$

D $5116\,kJ$.

10.   Atoms of nitrogen and element **X** form a bond in which the electrons are shared equally.

Element **X** could be

A   carbon

B   oxygen

C   chlorine

D   phosphorus.

11. Which line in the table represents the solid in which only van der Waals' forces are overcome when the substance melts?

	Melting point / °C	Electrical conduction of solid
A	714	non-conductor
B	98	conductor
C	660	conductor
D	44	non-conductor

12. Which of the following does **not** contain covalent bonds?

A Hydrogen gas

B Helium gas

C Nitrogen gas

D Solid sulphur

13. Which of the following structures is **never** found in compounds?

A Ionic

B Monatomic

C Covalent network

D Covalent molecular

14. In which of the following solvents is lithium chloride most likely to dissolve?

A Hexane

B Benzene

C Methanol

D Tetrachloromethane

15. A balloon contains 0·1 mol of oxygen gas, and 0·2 mol of carbon dioxide gas.

The total number of molecules in the balloon is approximately

A $6 \cdot 0 \times 10^{23}$

B $3 \cdot 6 \times 10^{23}$

C $2 \cdot 4 \times 10^{23}$

D $1 \cdot 8 \times 10^{23}$.

16. Which of the following pairs of gases occupy the same volume?

(Assume all measurements are made under the same conditions of temperature and pressure.)

A 2 g hydrogen and 14 g nitrogen

B 32 g methane and 88 g carbon dioxide

C 7 g carbon monoxide and 16 g oxygen

D 10 g hydrogen chloride and 10 g sulphur dioxide

17. $2NO(g) + O_2(g) \rightarrow 2NO_2(g)$

How many litres of nitrogen dioxide gas could theoretically be obtained in the reaction of 1 litre of nitrogen monoxide gas with 2 litres of oxygen gas?

(All volumes are measured under the same conditions of temperature and pressure.)

A 1

B 2

C 3

D 4

18. Which of the following hydrocarbons is **least** likely to be added to petrol to improve the efficiency of its burning?

A $CH_3CH_2CH_2CH_2CH_2CH_2CH_3$

B
$$\begin{array}{cc} CH_3 & CH_3 \\ | & | \\ CH_3CHCHCHCH_3 \\ | \\ CH_3 \end{array}$$

C
$$\begin{array}{c} CH_2 \\ H_2C \quad CH_2 \\ | \qquad | \\ H_2C \quad CH_2 \\ CH_2 \end{array}$$

D

[Turn over

19. Biogas is produced under anaerobic conditions by the fermentation of biological materials.

What is the main constituent of biogas?

A   Butane

B   Ethane

C   Methane

D   Propane

20. Which equation represents a reaction which takes place during reforming?

A $C_6H_{14} \rightarrow C_6H_6 + 4H_2$

B $C_4H_8 + H_2 \rightarrow C_4H_{10}$

C $C_2H_5OH \rightarrow C_2H_4 + H_2O$

D $C_8H_{18} \rightarrow C_4H_{10} + C_4H_8$

21. Which of the following is an isomer of 2-methyl pentane?

A
$$\begin{array}{cc} CH_3 & CH_3 \\ | & | \\ CH - CH \\ | & | \\ CH_3 & CH_3 \end{array}$$

B
$$\begin{array}{c} CH_3 \\ | \\ CH - CH_2 \\ | \qquad | \\ CH_3 \quad CH_2 - CH_3 \end{array}$$

C $CH_3 - CH_2 - CH_2 - \underset{\underset{CH_3}{|}}{\overset{\overset{CH_3}{|}}{CH}}$

D $CH_3 - \underset{\underset{}{|}}{\overset{\overset{CH_3}{|}}{CH}} - CH_2 - \underset{\underset{}{|}}{\overset{\overset{CH_3}{|}}{CH_2}}$

22. Which line in the table shows the correct functional group for each homologous series?

	Alkanoic acid	Alkanol	Alkanal
A	$-C\overset{\nearrow O}{\underset{\searrow H}{}}$	$-OH$	$-C\overset{\nearrow O}{\underset{\searrow OH}{}}$
B	$-C\overset{\nearrow O}{\underset{\searrow OH}{}}$	$-OH$	$-C\overset{\nearrow O}{\underset{\searrow H}{}}$
C	$-C\overset{\nearrow O}{\underset{\searrow OH}{}}$	$-C\overset{\nearrow O}{\underset{\searrow H}{}}$	$-OH$
D	$-OH$	$-C\overset{\nearrow O}{\underset{\searrow OH}{}}$	$-C\overset{\nearrow O}{\underset{\searrow H}{}}$

23. Hydrolysis of an ester gave an alkanol and an alkanoic acid both of which had the same molecular mass of 60.

The structure of the ester was

A
$$\begin{array}{c} \quad H \quad O \qquad\quad H \quad H \\ \quad | \quad || \qquad\quad | \quad | \\ H-C-C+O-C-C-H \\ \quad | \qquad\qquad\quad | \quad | \\ \quad H \qquad\qquad\quad H \quad H \end{array}$$

B
$$\begin{array}{c} \quad H \quad O \quad H \quad H \quad H \quad H \\ \quad | \quad || \quad | \quad | \quad | \quad | \\ H-C-C+O-C-C-C-H \\ \quad | \qquad\quad | \quad | \quad | \\ \quad H \qquad\quad H \quad H \quad H \end{array}$$

C
$$\begin{array}{c} \quad H \quad H \quad O \qquad\quad H \\ \quad | \quad | \quad || \qquad\quad | \\ H-C-C-C+O-C-H \\ \quad | \quad | \qquad\qquad\quad | \\ \quad H \quad H \qquad\qquad\quad H \end{array}$$

D
$$\begin{array}{c} \quad H \quad H \quad O \qquad\quad H \quad H \\ \quad | \quad | \quad || \qquad\quad | \quad | \\ H-C-C-C+O-C-C-H \\ \quad | \quad | \qquad\qquad\quad | \quad | \\ \quad H \quad H \qquad\qquad\quad H \quad H \end{array}$$

24. Which of the following statements about benzene is true?

(You may wish to refer to the Data Booklet.)

A Benzene has the same ratio of carbon to hydrogen as ethyne.

B Benzene reacts with copper(II) oxide more easily than ethanol.

C Benzene is more volatile than ethanal.

D Benzene undergoes addition reactions more readily than ethene.

25. Which of the following is a possible product of the reaction of propyne with bromine?

A
```
Br      H  H
  \     |  |
   C = C - C - H
  /        |
H          H
```

B
```
   Br Br  H
   |  |   |
Br-C- C - C - H
   |  |   |
   Br H   H
```

C
```
Br      Br H
  \     |  |
   C = C - C - H
  /        |
H          H
```

D
```
   Br H  Br
   |  |  |
H - C- C- C - H
   |  |  |
   Br H  Br
```

26. Which alcohol could be oxidised to a carboxylic acid?

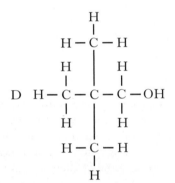

A
```
  H  H  H  H  H
  |  |  |  |  |
H-C--C--C--C--C-H
  |  |  |  |  |
  H  H  H  OH H
```

B
```
  H  H  H  H  H
  |  |  |  |  |
H-C--C--C--C--C-H
  |  |  |  |  |
  H  H  OH H  H
```

C
```
           H
           |
         H-C-H
  H        |    H  H
  |        |    |  |
H-C------- C -- C--C-H
  |        |    |  |
  H        OH   H  H
```

D
```
         H
         |
       H-C-H
  H      |    H
  |      |    |
H-C----- C -- C--OH
  |      |    |
  H      |    H
       H-C-H
         |
         H
```

27. What mixture of gases is known as synthesis gas?

A Methane and oxygen

B Carbon monoxide and oxygen

C Carbon dioxide and hydrogen

D Carbon monoxide and hydrogen

28. Which of the following substances does **not** have delocalised electrons?

A Aluminium

B Poly(ethyne)

C Poly(ethenol)

D Carbon (graphite)

[Turn over

29. The arrangement of amino acids in a peptide is

$$Z-X-W-V-Y$$

where the letters V, W, X, Y and Z represent amino acids.

On partial hydrolysis of the peptide, which of the following sets of dipeptides is possible?

A   V-Y, Z-X, W-Y, X-W

B   Z-X, V-Y, W-V, X-W

C   Z-X, X-V, W-V, V-Y

D   X-W, X-Z, Z-W, Y-V

30. Aluminium reacts with oxygen to form aluminium oxide.

$$2Al(s) + 1\frac{1}{2}O_2(g) \rightarrow Al_2O_3(s) \quad \Delta H = -1670\,kJ\,mol^{-1}$$

What is the enthalpy of combustion of aluminium in $kJ\,mol^{-1}$?

A   −835

B   −1113

C   −1670

D   +1670

31. A few drops of concentrated sulphuric acid were added to a mixture of 0·1 mol of methanol and 0·2 mol of ethanoic acid. Even after a considerable time, the reaction mixture was found to contain some of each reactant.

Which of the following is the best explanation for the incomplete reaction?

A   The temperature was too low.

B   An equilibrium mixture was formed.

C   Insufficient methanol was used.

D   Insufficient ethanoic acid was used.

32. Which line in the table applies correctly to the use of a catalyst in a chemical reaction?

	Position of equilibrium	Effect on value of ΔH
A	Moved to right	Decreased
B	Unaffected	Increased
C	Moved to left	Unaffected
D	Unaffected	Unaffected

33. The hypochlorite ion, $ClO^-(aq)$, produced in the reaction shown, is used as a bleach.

$$Cl_2(g) + H_2O(\ell) \rightleftharpoons 2H^+(aq) + ClO^-(aq) + Cl^-(aq)$$

The concentration of ClO^- ions could be increased by the addition of

A   solid potassium hydroxide

B   concentrated hydrochloric acid solution

C   solid sodium chloride

D   solid potassium sulphate.

34. A solution of hydrochloric acid with a pH of 6 and another of sodium hydroxide with a pH of 8 are each diluted by a factor of 100.

After dilution, when tested using pH indicator paper,

A   the pH of the acid drops by 2

B   the pH of the alkali rises by 2

C   the pH of the acid equals that of the alkali

D   the pH of the acid is 2 below that of the alkali.

35. Equal volumes of four $1\,mol\,l^{-1}$ solutions were compared.

Which of the following $1\,mol\,l^{-1}$ solutions contains the fewest ions?

A   Hydrochloric acid

B   Ethanoic acid

C   Sodium chloride

D   Sodium hydroxide

36. Equal volumes of 0.1 mol l^{-1} solutions of sodium hydroxide and propanoic acid were mixed together.

The pH of the resulting solution could be

A　3

B　5

C　7

D　9.

37. During a redox process in acid solution, iodate ions, $IO_3^-(aq)$, are converted into iodine, $I_2(aq)$.

$$IO_3^-(aq) \rightarrow I_2(aq)$$

The numbers of $H^+(aq)$ and $H_2O(\ell)$ required to balance the ion-electron equation for the formation of 1 mol of $I_2(aq)$ are, respectively

A　3 and 6

B　6 and 3

C　6 and 12

D　12 and 6.

38. One mole of metal would be deposited by passing $96\,500$ C through a solution of

A　silver(I) nitrate

B　gold(III) nitrate

C　nickel(II) nitrate

D　copper(II) nitrate.

39. $^2_1H + {}^3_1H \rightarrow {}^4_2He + {}^1_0n$

The above process represents

A　nuclear fission

B　nuclear fusion

C　proton capture

D　neutron capture.

40. Naturally occurring nitrogen consists of two isotopes ^{14}N and ^{15}N.

How many different nitrogen molecules will occur in the air?

A　1

B　2

C　3

D　4

Candidates are reminded that the answer sheet MUST be returned INSIDE the front cover of this answer book.

[Turn over

[BLANK PAGE]

Marks

SECTION B

All answers must be written clearly and legibly in ink.

1. The elements lithium to neon make up the second period of the Periodic Table.

(a) Name an element from the second period that exists as a covalent network.

Carbon

1

(b) Why do the atoms decrease in size from lithium to neon?

protons increase as you go across the period.

1

(2)

[Turn over

Marks

2. Copper(II) carbonate reacts with dilute hydrochloric acid as shown.

$$CuCO_3(s) + 2HCl(aq) \rightarrow CuCl_2(aq) + H_2O(\ell) + CO_2(g)$$

A student used the apparatus shown below to follow the progress of the reaction.

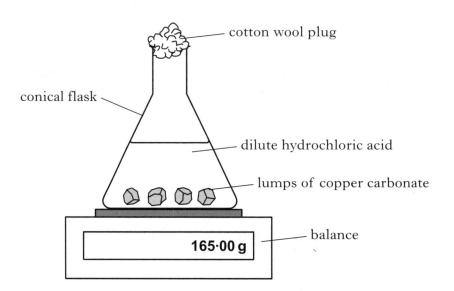

cotton wool plug

conical flask

dilute hydrochloric acid

lumps of copper carbonate

165·00 g

balance

(a) Suggest why a cotton wool plug is placed in the mouth of the conical flask.

to prevent loss of any solution, as they spray

1

Marks

2. (continued)

(*b*) The experiment was carried out using $0.50\,g$ samples of both pure and impure copper(II) carbonate. The graph below shows the results obtained.

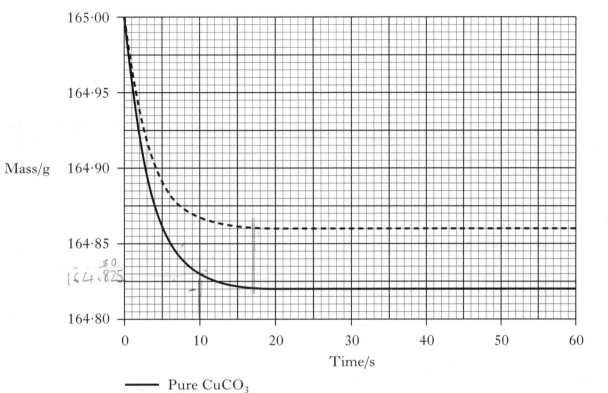

Mass/g

——— Pure $CuCO_3$

- - - - Impure $CuCO_3$

(i) For the sample of pure copper(II) carbonate, calculate the average reaction rate, in $g\,s^{-1}$, over the first 10 seconds.

$$165 - 164.830 \qquad \frac{0.17}{10}$$

$$= 0.017\,g\,s^{-1}$$

1

(ii) Calculate the mass, in grams, of copper(II) carbonate present in the impure sample.

Show your working clearly.

0.50

1

(3)

Marks

3. Ethanol, C_2H_5OH, can be used as a fuel in some camping stoves.

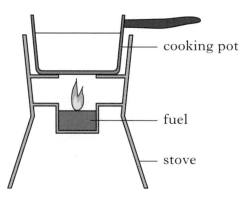

cooking pot

fuel

stove

(a) The enthalpy of combustion of ethanol is $-1367\,kJ\,mol^{-1}$.

Using this value, calculate the number of moles of ethanol required to raise the temperature of $500\,g$ of water from $18\,^\circ C$ to $100\,^\circ C$.

Show your working clearly.

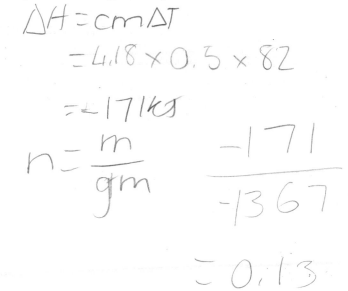

$$\Delta H = cm\Delta T$$
$$= 4.18 \times 0.5 \times 82$$
$$= -171\,kJ$$
$$n = \frac{m}{gfm} \qquad \frac{-171}{-1367}$$
$$= 0.13$$

2

(b) Suggest **two** reasons why less energy is obtained from burning ethanol in the camping stove than is predicted from its enthalpy of combustion.

heat is lost to surroundings
incomplete combustion
of OI ERI

2

(4)

Marks

4. Phosphorus-32 and strontium-89 are two radioisotopes used to study how far mosquitoes travel.

   (a) Strontium-89 decays by emission of a beta particle.

   Complete the nuclear equation for the decay of strontium-89.

   $$^{89}_{38}Sr \rightarrow \ ^{89}_{39}Y \ \ ^{0}_{-1}e$$ 1

   (b) In an experiment, 10 g of strontium-89 chloride was added to a sugar solution used to feed mosquitoes.

   (i) The strontium-89 chloride solution was fed to the mosquitoes in a laboratory at 20 °C. When the mosquitoes were released, the outdoor temperature was found to be 35 °C.

   What effect would the increase in temperature have on the half-life of the strontium-89?

   no effect

   1

   (ii) Calculate the mass, in grams, of strontium-89 present in the 10 g sample of strontium-89 chloride, $SrCl_2$.

   1

   (c) A mosquito fed on a solution containing phosphorus-32 is released.

   Phosphorus-32 has a half-life of 14 days.

   When the mosquito is recaptured 28 days later, what fraction of the phosphorus-32 will remain?

   $$\frac{1}{2}$$

   $$\frac{1}{4} = 0.25\%$$

   1

   (4)

5. The concentration of ethanol in a person's breath can be determined by measuring the voltage produced in an electrochemical cell.

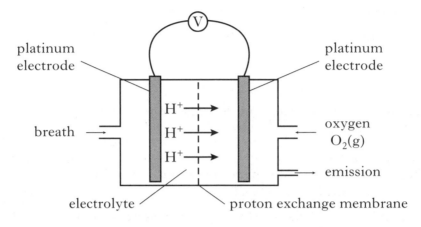

Different ethanol vapour concentrations produce different voltages as is shown in the graph below.

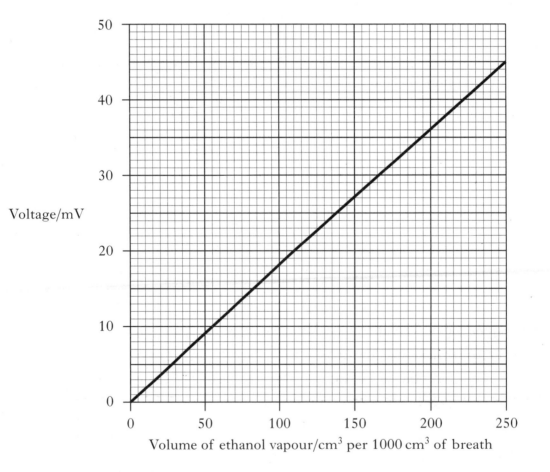

Marks

5.   (continued)

(*a*) Calculate the number of ethanol molecules in $1000 \, cm^3$ of breath when a voltage of $20 \, mV$ was recorded.

(Take the molar volume of ethanol vapour to be 24 litres mol^{-1}.)

Show your working clearly.

2

(*b*) The ion-electron equations for the reduction and oxidation reactions occurring in the cell are shown below.

$$O_2 + 4H^+ + 4e^- \rightarrow 2H_2O$$

$$CH_3CH_2OH + H_2O \rightarrow CH_3COOH + 4H^+ + 4e^-$$

Write the overall redox equation for the reaction taking place.

$O_2 + CH_3CH_2OH \longrightarrow H_2O + CHCOOH$

1

(*c*) Platinum metal acts as a heterogeneous catalyst in this reaction.

What is meant by a **heterogeneous catalyst**?

catalyst when is in a different state the reactant.

1

(4)

Marks

6. Compounds containing sulphur occur widely in nature.

(*a*) The compound dimethyldisulphide, $CH_3S_2CH_3$, is present in garlic and onions.

Draw a full structural formula for this compound.

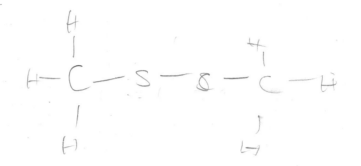

1

(*b*) Hydrogen sulphide, H_2S, formed by the decomposition of proteins, can cause an unpleasant odour in water supplies.

(i) Chlorine, added to the water, removes the hydrogen sulphide.

The equation for the reaction taking place is

$$4Cl_2(aq) + H_2S(aq) + 4H_2O(\ell) \rightarrow SO_4^{2-}(aq) + 10H^+(aq) + 8Cl^-(aq)$$

An average of $29.4\,cm^3$ of $0.010\,mol\,l^{-1}$ chlorine solution was required to react completely with a $50.0\,cm^3$ sample of water.

Calculate the hydrogen sulphide concentration, in $mol\,l^{-1}$, present in the water sample.

Show your working clearly.

$$\frac{C_1 V_1}{N_1} = \frac{C_2 V_2}{N_2}$$

$$\frac{0.010 \times \overset{0.0294}{294}}{4} = \frac{C_2 \times 0.05}{1}$$

$$0.0735 = 0.05C$$

$$C = 1.47$$

$$7.35 \times 10^{-5} = 0.05C$$

$$C = 1.47 \times 10^{-3}$$

2

Marks

6. (b) (continued)

   (ii) Liquid hydrogen sulphide has a boiling point of −60 °C.

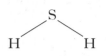

   Explain clearly why hydrogen sulphide is a gas at room temperature. In your answer, you should name the intermolecular forces involved and indicate how they arise.

2

(5)

[Turn over

Marks

7. Polyurethanes are polymers that are widely used in industry. They are produced by the reaction of diisocyanates with diols.

(*a*) The structure of one such diisocyanate is shown below.

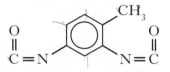

The molecular formula for this compound can be written as $C_wH_xN_yO_z$.

Give the values for **w**, **x**, **y** and **z**.

w = 9   **x** = 6   **y** = 2   **z** = 2

1

(*b*) Lycra is a polyurethane polymer made by the polymerisation of 2-diisocyanatoethane with ethane-1,2-diol. The reaction to form the repeating unit is shown.

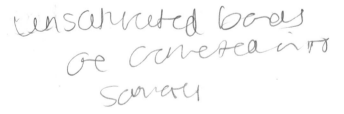

Why is this polymerisation described as an addition reaction?

unsaturated bonds
or converted into
sorgey

1

Marks

7. **(continued)**

(*c*) Lycra is strong because of the hydrogen bonding between neighbouring polymer chains.

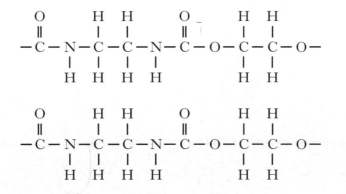

Draw a dotted line to show a hydrogen bond between the polymer chains above.

1

(3)

[**Turn over**

Marks

8. Aspartame is an artificial sweetener which has the structure shown below.

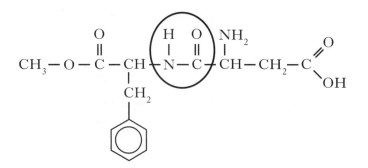

(a) Name the functional group circled.

amide link

1

(b) In the stomach, aspartame is hydrolysed by acid to produce methanol and two amino acids, phenylalanine and aspartic acid.

Two of the products of the hydrolysis of aspartame are shown below.

$CH_3 - OH$

methanol

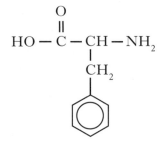

phenylalanine

Draw a structural formula for aspartic acid.

1

Marks

8.   **(continued)**

(*c*)   The body cannot make all the amino acids it requires and is dependent on protein in the diet for the supply of certain amino acids.

What term is used to describe the amino acids the body cannot make?

essential amino acid

1

(*d*)   To investigate this hydrolysis reaction in the lab, the apparatus shown below is set up. The extent of hydrolysis at a given temperature can be determined by measuring the quantity of methanol produced.

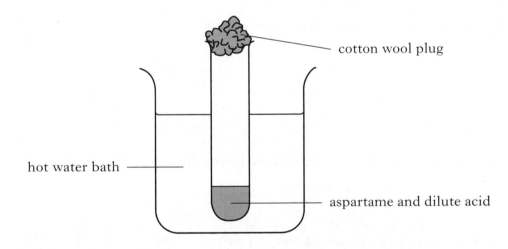

What improvement could be made to the **apparatus** to reduce the loss of methanol by evaporation?

wet paper towel on top of test tube to condense gas

1
(4)

DO NOT
WRITE
IN THIS
MARGIN

Marks

9.   A fatty acid is a long chain carboxylic acid.

Examples of fatty acids are shown in the table below.

Common name	Systematic name	Structure
stearic acid	octadecanoic acid	$CH_3(CH_2)_{16}COOH$
oleic acid	octadec-9-enoic acid	$CH_3(CH_2)_7CH=CH(CH_2)_7COOH$
linoleic acid	octadec-9,12-dienoic acid	$CH_3(CH_2)_4CH=CHCH_2CH=CH(CH_2)_7COOH$
linolenic acid		$CH_3CH_2CH=CHCH_2CH=CHCH_2CH=CH(CH_2)_7COOH$

(a)   Describe a chemical test, with the expected result, that could be used to distinguish between stearic and oleic acids.

Oleic decaluri bromi sa

1

(b)   What is the systematic name for linolenic acid?

octadec   acid

1

(c)   Stearic acid can be reacted with sodium hydroxide solution to make a soap.

The structure of the soap is shown.

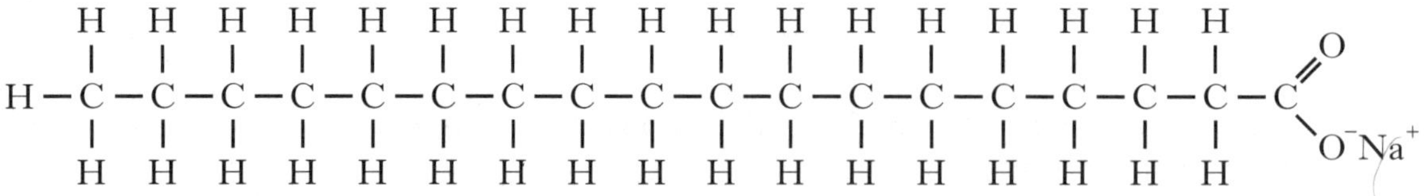

One part of the soap molecule is soluble in fat and the other part is soluble in water.

Circle the part of the soap molecule which is soluble in water.

1

(3)

$O^- Na^+$

Marks

10. Nitrogen and compounds containing nitrogen are widely used in industry.

(*a*) From which raw material is nitrogen obtained?

air

1

(*b*) In industry, methanamide, $HCONH_2$, is produced from the ester shown below.

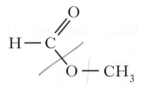

Name the ester.

methyl methanoal

1

(*c*) In the lab, methanamide can be prepared by the reaction of methanoic acid with ammonia.

$$HCOOH + NH_3 \rightleftharpoons HCONH_2 + H_2O$$

methanoic   methanamide
acid

When 1·38 g of methanoic acid was reacted with excess ammonia, 0·945 g of methanamide was produced.

Calculate the percentage yield of methanamide.

Show your working clearly.

1 mol — 1 mol

46 g — 45 g

1.38 g ×x

46x — 62.1

= x — 1.35

HCOOH
2
32
12
46

HCONH₂
3
14
16
12

$x = \frac{A}{T} \times 100 =$

$= \frac{0.945}{1.35} \times 100$

= 70%

2

(4)

Marks

11. The element boron forms many useful compounds.

(a) Borane (BH_3) is used to synthesize alcohols from alkenes.

The reaction occurs in two stages

Stage 1 Addition Reaction

The boron atom bonds to the carbon atom of the double bond which already has the most hydrogens **directly** attached to it.

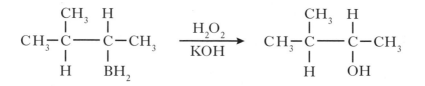

Stage 2 Oxidation Reaction

The organoborane compound is oxidised to form the alcohol.

$$\begin{array}{c} CH_3\ H \\ |\quad | \\ CH_3-C-C-CH_3 \\ |\quad | \\ H\quad BH_2 \end{array} \xrightarrow[\text{KOH}]{H_2O_2} \begin{array}{c} CH_3\ H \\ |\quad | \\ CH_3-C-C-CH_3 \\ |\quad | \\ H\quad OH \end{array}$$

(i) Name the alcohol produced in Stage 2.

ethan

1

(ii) Draw a structural formula for the **alcohol** which would be formed from the alkene shown below.

$$\begin{array}{c} \quad\quad\quad\quad CH_3\ H \\ \quad\quad\quad\quad |\quad | \\ CH_3-CH_2-CH_2-C=C-H \end{array}$$

1

Marks

11. **(continued)**

(b) The compound diborane (B_2H_6) is used as a rocket fuel.

(i) It can be prepared as shown.

$$BF_3 + \quad NaBH_4 \rightarrow \quad B_2H_6 + \quad NaBF_4$$

Balance this equation.

1

(ii) The equation for the combustion of diborane is shown below.

$$B_2H_6(g) + 3O_2(g) \rightarrow B_2O_3(s) + 3H_2O(\ell)$$

Calculate the enthalpy of combustion of diborane (B_2H_6) using the following data.

$$2B(s) + 3H_2(g) \rightarrow B_2H_6(g) \quad \Delta H = 36 \, \text{kJ mol}^{-1}$$
$$H_2(g) + \tfrac{1}{2}O_2(g) \rightarrow H_2O(\ell) \quad \Delta H = -286 \, \text{kJ mol}^{-1}$$
$$2B(s) + 1\tfrac{1}{2}O_2(g) \rightarrow B_2O_3(s) \quad \Delta H = -1274 \, \text{kJ mol}^{-1}$$

$\Delta H =$

2

(c) Diborane can be used to manufacture pentaborane (B_5H_9).

Pentaborane was also considered for use as a rocket fuel because its enthalpy of combustion is $-9037 \, \text{kJ mol}^{-1}$.

Calculate the energy released, in kJ, when 1 kilogram of pentaborane is completely burned.

1

(6)

Marks

12. In the PPA experiment **Factors Affecting Enzyme Activity**, a student investigated the effect of pH on the activity of the enzyme catalase contained in potato discs.

The following apparatus was set up and left for 3 minutes.

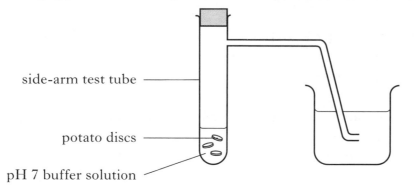

side-arm test tube

potato discs

pH 7 buffer solution

(a) Why must the buffer/potato disc mixture be left for 3 minutes before continuing the experiment?

1

(b) Which chemical must then be added to the test tube to investigate the enzyme activity at this pH?

1

(c) The activity of the enzyme was measured at several different pH values by counting the number of bubbles produced in a given time.

Why were no bubbles produced at pH 1?

1

(3)

DO NOT WRITE IN THIS MARGIN

Marks

13. Fluorine is an extremely reactive element. Its compounds are found in a range of products.

(*a*) Fluorine gas can be produced by electrolysis.

The ion-electron equation for the production of fluorine gas is:

$$2F^{-}(\ell) \rightarrow F_2(g) + 2e^{-}$$

Calculate the mass, in grams, of fluorine produced when a steady current of 5·0 A is passed through the solution for 32 minutes.

Show your working clearly. 2

[Turn over

Marks

13. **(continued)**

(b) Tetrafluoroethene, C_2F_4, is produced in industry by a series of reactions.

The final reaction in its manufacture is shown below.

$$2CHClF_2(g) \rightleftharpoons C_2F_4(g) + 2HCl(g)$$

(i) The graph shows the variation in the concentration of C_2F_4 formed at equilibrium as temperature is increased.

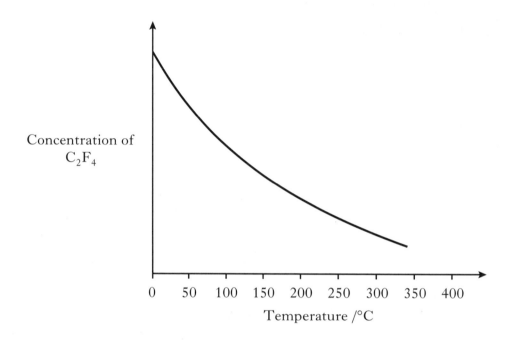

What conclusion can be drawn about the enthalpy change for the formation of tetrafluoroethene?

1

Marks

13. **(b)** **(continued)**

(ii) Sketch a graph to show how the concentration of C_2F_4 formed at equilibrium would vary with increasing pressure.

(An additional graph, if required, can be found on *Page thirty-eight.*)

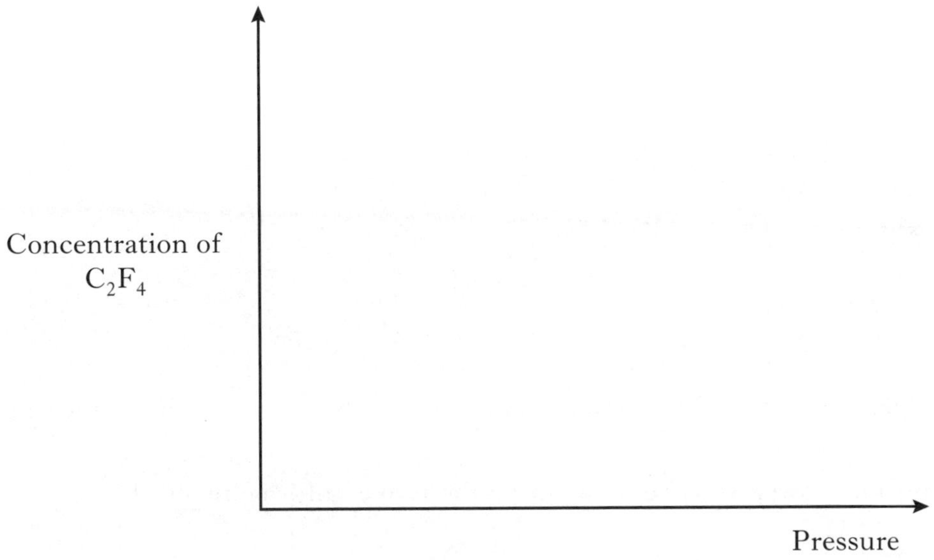

Concentration of
C_2F_4

Pressure

1

(c) Hydrochlorofluorocarbons are used as replacements for chlorofluorocarbons, CFCs.

What environmental problem is associated with the extensive use of CFCs?

1

(5)

[Turn over

Marks

14. Ammonium nitrate (NH_4NO_3) is widely used as a fertiliser.

    (a) An ammonium nitrate solution has a pH of 5.

       (i) Calculate the concentration, in $mol\,l^{-1}$, of $H^+(aq)$ ions in the solution.

       1

       (ii) **Explain clearly** why ammonium nitrate dissolves in water to produce an acidic solution.

       In your answer, you should mention the **two** equilibria involved.

       2

DO NOT WRITE IN THIS MARGIN

Marks

14. (continued)

(*b*) Ammonium nitrate must be stored and transported carefully as it can decompose according to the equation shown below:

$$2NH_4NO_3(s) \rightarrow 2N_2(g) + O_2(g) + 4H_2O(g)$$

In addition to being very exothermic, suggest another reason why the decomposition of ammonium nitrate can result in an explosion. **1**

(4)

[Turn over

Marks

15. Hydrogen gas can be produced in a variety of ways.

(a) Hydrogen can be produced in the lab from dilute sulphuric acid. The apparatus shown below can be used to determine the quantity of electrical charge required to form one mole of hydrogen gas.

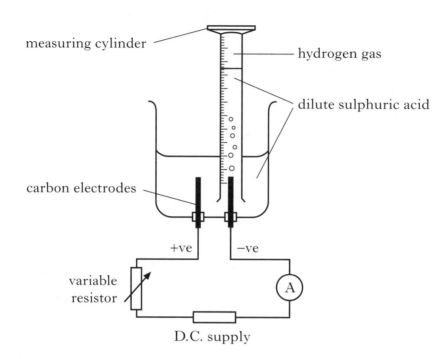

(i) Why is a variable resistor used?

1

(ii) In addition to measuring the volume of hydrogen gas, which **two** other measurements must be made?

1

Marks

15. **(continued)**

(b) The sulphur-iodine cycle is an industrial process used to manufacture hydrogen.

There are three steps in the sulphur-iodine cycle.

Step 1: $I_2 + SO_2 + 2H_2O \rightarrow 2HI + H_2SO_4$

Step 2: $2HI \rightarrow I_2 + H_2$

Step 3: $H_2SO_4 \rightarrow SO_2 + H_2O + \frac{1}{2}O_2$

(i) Why does step 3 help to reduce the cost of manufacturing hydrogen?

1

(ii) What is the overall equation for the sulphur-iodine cycle?

1

(4)

[Turn over

Marks

16. In alkane molecules, the chains of carbon atoms are very flexible. The molecules can twist at any carbon-to-carbon bond.

    Butane is a typical alkane.

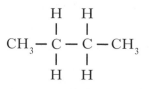

    The diagram below shows butane twisting about the central carbon-to-carbon bond.

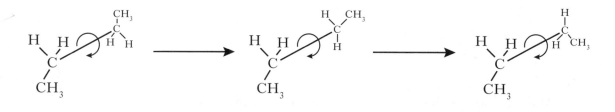

    Newman projections are special diagrams used to show the relative position of the different atoms in a molecule.

    In the diagram below, a Newman projection has been drawn showing the relative position of the atoms in the butane molecule below.

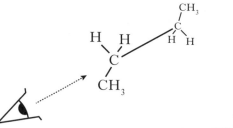 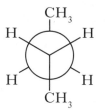

Newman projection

   (a) Complete the Newman projection diagram for the butane molecule in the position shown below.

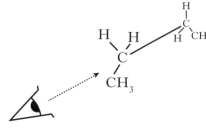

Newman projection   1

Marks

16.   **(continued)**

(*b*)   Name the alkane molecule represented by the Newman projection shown below.

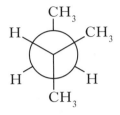

1

(2)

[*END OF QUESTION PAPER*]

ADDITIONAL GRAPH FOR USE IN QUESTION 13(*b*)(ii)

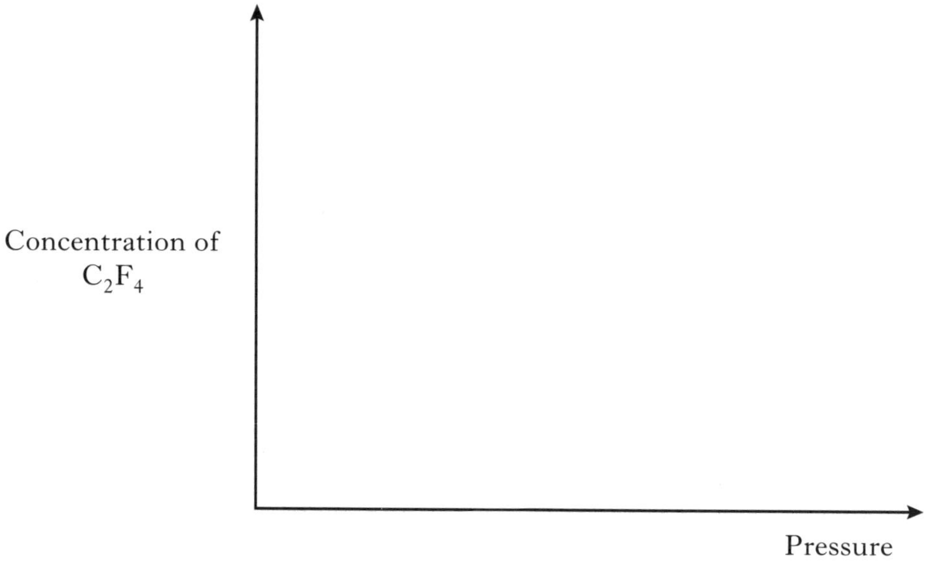

Concentration of C_2F_4

Pressure

ADDITIONAL SPACE FOR ANSWERS

[BLANK PAGE]

HIGHER | ANSWER SECTION

SQA HIGHER
CHEMISTRY 2008–2012

CHEMISTRY HIGHER 2008

SECTION A

1. D	11. C	21. D	31. B
2. B	12. D	22. A	32. B
3. A	13. A	23. A	33. C
4. D	14. C	24. B	34. A
5. A	15. B	25. A	35. C
6. D	16. C	26. B	36. D
7. D	17. D	27. C	37. D
8. C	18. B	28. B	38. A
9. A	19. C	29. B	39. C
10. D	20. B	30. C	40. C

SECTION B

1. CO_2 covalent molecular **or** discrete covalent
    SiO_2 covalent network **or** covalent lattice

2. (a) esters

    (b) *Any one from:*
    they react with hydrogen (or are hydrogenated)
    they become (more) saturated (or less unsaturated)
    they have fewer double bonds (or more single bonds)
    the double bonds are broken

    (c) *Any one from:*
    as an energy source (or more concentrated energy source than carbohydrates)
    provide essential fatty acids
    carry oil soluble vitamins
    good for health with reason given, eg lowers cholesterol

3. (a) a certain volume of KI solution was measured out and the volume made up to 25 cm³ with water (and this was repeated)
    or 20 cm³ KI solution added to 5 cm³ of water;
    15 cm³ KI solution to 10 cm³ of water, etc

    (b) 23·3 s

4. (a) synthesis gas

    (b)
    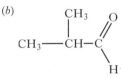

    (c) (i) any mention of silver being formed (deposited)

        (ii) in a water bath **or** no naked flames

    (d) primary

5. (a) $^{0}_{-1}e$

        or beta (particle)

    (b) (i)
    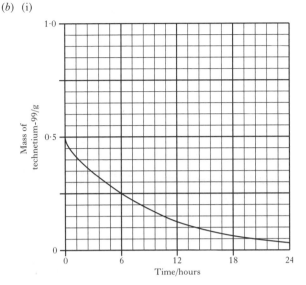

        (ii) short half-life
        or little will remain in the body after a short time **or** not as ionising (as other radiation)

6. (a) (i) benzene does not (rapidly) decolourise bromine (solution or water)

        (ii) *Any two from:*
        mention of delocalised electrons
        C to C bonds are all of equal length
        planar molecule
        bond angles of 120°

    (b) reforming

    (c) Increases efficiency of burning
    or cuts down auto ignition (or knocking)

7. (a) depletion of the ozone layer

    (b) 615 litres

8. (a) $2B_2O_3 + 7C \rightarrow B_4C + 6CO$

    (b) **X**: any workable method of producing CO_2 with calcium carbonate and dilute hydrochloric acid labelled
    Y: any workable method of removing CO_2 with chemical labelled, eg sodium hydroxide solution, lime water, alkali

    (c) incomplete combustion

9. (a) *Both points required:*
    weak van der Waals' forces
    result of instantaneous dipoles caused by movement of electrons

    (b) (i) to saturate the (porous) carbon rods with (hydrogen) gas
    or the gas trapped in the carbon rods leads to an error

        (ii) 0·00187 g

    (c) 10^{-13} mol l^{-1}

10. (*a*) any suitable indication of point at which curves start to level off on concentration axis, eg by a vertical line or arrow

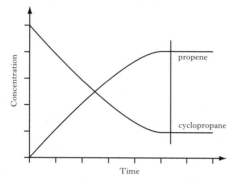

(*b*) the ratio of moles of reactant (gas): moles of product gas is 1:1

or the number of (gaseous) molecules is the same on both sides of the equation

(*c*)

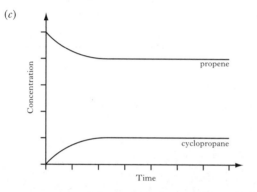

11. (*a*) alcohols do not contain OH⁻ hydroxide ions **or** the OH in alcohols is not ionic

(*b*) *Both points:*
potassium iodide contains iodide ions
the iodine molecules form the blue/black colour with starch

12. (*a*) the **shape** of the reactant (substrate) molecule must fit the enzyme

(*b*) $-202 \cdot 6$ kJ mol^{-1}

13. (*a*) hydroxyl

(*b*) (i) biopol or other biodegradable polymer, eg starch, cellulose, protein, poly(ethenol)

(ii) The reactants are allowed to be used up (and then the products are removed) before fresh reactants are added (or the process is restarted).

(iii)

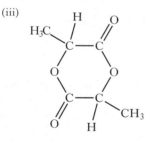

14. (*a*) homogeneous

(*b*) (i) no effect
(ii) -152 kJ

15. (*a*) (i) HO–CH$_2$–CH$_2$–OH

(ii) sodium chloride will cause rusting

(*b*) butane

16. (*a*)

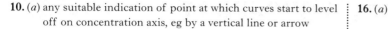

(*b*) methanal **or** 2,2-dimethylpropanal

(*c*) water is not a product of the reaction **or** no small molecule produced **or** it is an addition reaction

17. (*a*) $MnO_4^-(aq) + 8H^+(aq) + 5e^- \rightarrow Mn^{2+}(aq) + 4H_2O(l)$

(*b*) there is a colour change from colourless to purple

(*c*) (i) first titre is a rough (or approximate) result
(ii) $0 \cdot 05$

CHEMISTRY HIGHER 2009

SECTION A

1. D	11. C	21. A	31. D
2. C	12. C	22. C	32. A
3. C	13. D	23. B	33. D
4. A	14. A	24. A	34. B
5. D	15. D	25. B	35. B
6. A	16. C	26. B	36. D
7. B	17. D	27. C	37. D
8. D	18. C	28. A	38. B
9. A	19. B	29. C	39. A
10. B	20. D	30. A	40. C

SECTION B

1. (a) increases (or gets bigger or rises)

   (b) more energy is needed to remove the electron from a full shell (or complete shell or noble gas shell) **or**
   an electron is being removed from an energy level closer to the nucleus **or**
   there is a greater nuclear pull on the electron being removed **or**
   second energy level is nearer the nucleus **or**
   second energy level is full (or complete), etc.

   (c) forces of attraction **between** molecules (or intermolecular forces or van der Waals' forces) increase **or** energy needed to separate the molecules increases

2. (a) **x = 7 y = 8**

   (b) non-polluting **or** no greenhouse gases **or** no carbon dioxide produced **or** burns to produce water only **or** a cleaner fuel, etc.

   (c) renewable (or not a finite resource or carbon-neutral)

3. (a) (i) ratio of oxygen:hydrogen atoms increased (or ratio of hydrogen:oxygen atoms decreased) **or** removal of hydrogen

   (ii) orange to green (or blue/green)

   (b) (i) Any mention of separate layer **or** any mention of (ester) smell

   (ii)
   $$CH_3\text{-}CH_2\text{-}\overset{\overset{\displaystyle O}{\|}}{C}\text{-}O\text{-}CH_2\text{-}CH_2\text{-}CH_3$$

4. (a) absorbs (harmful) UV radiation **or** reduces (or stops) UV radiation reaching earth

   (b) heterogeneous

   (c) $3O_2 \rightarrow 2O_3$
   1 mol → 2/3 mol
   1 mol = 6×10^{23} molecules
   2/3 mol = 4×10^{23} molecules

5. (a) (i) amino acids

   (ii) breaking up (bonds in) a molecule by the addition of (the elements from) water

   (b) (i) ester

   (ii) functional groups are only at the end of the monomers (or monomers have only two functional groups) **or** no chance of cross-linking

6. (a) (i) initial and final temperature (or temperature range) volume (or mass) of water

   (ii) 0·370 g ⟷ 3·86 kJ

   32 g (1 mol CH_3OH) ⟷ $\dfrac{3\cdot86\times32}{0\cdot370}$

   = − 333·8 kJ mol^{-1}

   (b) complete combustion (or incomplete combustion in lab method) **or** richer supply of oxygen (or burns in air in lab method) **or** no evaporation of methanol

7. (a) use an (upturned) measuring cylinder (or graduated tube) filled with water **or** collect gas over water **or** correct diagram

   (b) mass (or weight) **or** pH **or** concentration of acid **or** conductivity

8. (a) continuous

   (b) (i) yield decreases at high temperature;
   idea that equilibrium moves to the left (or to reactant side) at high temperature **or**
   corresponding explanation based on higher yield at lower temperatures

   (ii) idea that the formation of ammonia decreases the number of molecules (or reduces the pressure);
   idea that high pressure causes the equilibrium position to move to right (or product side) **or**
   high pressure favours the reaction that reduces the pressure

   (c) 1 mol N_2 → 2 mol NH_3

   28 g → 34 g

   500 kg → $\dfrac{500 \times 28}{34}$ = 607 kg

   % yield = $\dfrac{\text{actual}}{\text{theoretical}} \times 100 = \dfrac{405}{607} \times 100 = 66\cdot7\%$

   or

   no. of moles of $N_2 = \dfrac{500\ 000}{28}$ = 17 860 mol

   no. of moles of NH_3 = 35 720 mol = 607 kg

9. (a)

   primary   secondary   tertiary

   or

   Primary: hydroxyl group attached to C attached to two H atoms (or hydroxyl group attached to C attached to one C atom)

   Secondary: hydroxyl group attached to C attached to one H atom (or hydroxyl group attached to C attached to two C atoms)

   Tertiary: hydroxyl group attached to C attached to no H atoms (or hydroxyl group attached to C attached to three C atoms)

   (b) addition

   (c) pentan-3-one

10. (a) neutralisation

(b)
$$
\begin{array}{c}
O \\
\parallel \\
C-OH \\
\mid \\
H-C-OH \\
\mid \\
H-C-OH \\
\mid \\
C-OH \\
\parallel \\
O
\end{array}
$$

(c) 1 mol $C_4H_6O_6$ $\rightarrow$ 2 mol CO_2 = 48 litre

150 g $\rightarrow$ 48 litre

$\dfrac{150 \times 0 \cdot 105}{48}$ g $\rightarrow$ 0·105 litre

= 0·33 g

mass in 1 sweet = 0·0165 g

or no. of moles of $CO_2 = \dfrac{0 \cdot 105}{24} = 0 \cdot 0044$ mol

no. of moles of $C_4H_6O_6 = 0 \cdot 0022$ mol

$= 0 \cdot 0022 \times 150 = 0 \cdot 33$ g

mass in 1 sweet = 0·0165 g

11. (a) more collisions with energy greater or equal to E_a **or**
more collisions leading to an activated complex **or**
correct energy distribution diagram

(b) the outer electron in potassium is further from the nucleus
or
the outer electron is in a higher (or the fourth) energy level
or
the inner shells screen (or shield) the outer electron from the (pull of the) nucleus **or**
corresponding explanation based on chlorine

12. (a) no. of moles sulphuric acid = 0·05 × 0·01
= 0·0005

28 cm^3 $\longleftrightarrow$ 0·0005 mol

1 litre $\longleftrightarrow$ $\dfrac{0 \cdot 0005 \times 1000}{28} = 0 \cdot 018$ mol l^{-1}

or 0·01 × 50 = c × 28 c = 0·018 mol l^{-1}

(b) ions in barium sulphate are not free to move;
water contains few ions (or is made up mainly of molecules)

13. (a) add an ammeter;
add a variable resistor (or constant current supply)

(b) no. of moles Ag = $\dfrac{0 \cdot 365}{107 \cdot 9} = 0 \cdot 0034$

2 mol Ag $\longleftrightarrow$ 1 mol Cu
no. of moles Cu = 0·0017
mass of Cu = 0·0017 × 63·5 = 0·107 g
or
Ag 107·9 g $\longleftrightarrow$ 96 500 C

0·365 g $\longleftrightarrow$ $\dfrac{96500 \times 0 \cdot 365}{107 \cdot 9} = 326 \cdot 4$ C

Cu 2 × 96 500 C $\longleftrightarrow$ 63·5 g

326·4 C $\longleftrightarrow$ $\dfrac{63 \cdot 5 \times 326 \cdot 4}{2 \times 96500} = 0 \cdot 107$ g

14. (a) Exp. 2 lower; Exp. 3 same

(b) (i) pH = 1·0 $[H^+] = 1 \times 10^{-1}$ mol l^{-1}

$[OH^-] = \dfrac{1 \times 10^{-14}}{1 \times 10^{-1}} = 1 \times 10^{-13}$ mol l^{-1}

(ii) more chlorine atoms, the greater the strength of acid;
strength of acid related to degree of dissociation

15. (a) $Al_4C_3 + 12H_2O \rightarrow 4Al(OH)_3 + 3CH_4$

(b) $SiO_2(s) + 2H_2O(l) \rightarrow SiH_4(g) + 2O_2(g)$ +1517 kJ

Si(s) + O_2(g) $\rightarrow$ SiO_2(s) − 911 kJ

$2H_2$(g) + O_2(g) $\rightarrow$ $2H_2O$(g) − 572 kJ

addition = 34 kJ mol^{-1}

16. (a) ^{227}Th $\rightarrow$ ^{223}Th + α (or alpha or He)

(b) range of travel is short

(c) idea of 3 half-lives;
initial mass = 0·48 g

17. (a) 1·17 (or 7:6)

(b) (i) left to right 3, 2, 1

(ii) but-2-ene

18. (a) $2S_2O_3^{2-}$(aq) $\rightarrow$ $S_4O_6^{2-}$(aq) + 2e$^-$

(b) starch (solution)

(c) no. of moles of $S_2O_3^{2-}$ (aq) = 0·0504 × 0·10
= 0·00504

mole ratio 2:5
no. of moles of CO = 0·0125
or
no. of moles of $S_2O_3^{2-}$ (aq) = 0·504 × 0·10
= 0·00504

moles of iodine : thiosulphate is 1:2
moles of iodine = 0·0025
moles of CO : iodine is 5:1
moles of CO = 0·0125

CHEMISTRY HIGHER
2010

SECTION A

1.	B	11.	B	21.	C	31.	B
2.	C	12.	A	22.	C	32.	A
3.	C	13.	B	23.	A	33.	B
4.	B	14.	D	24.	C	34.	C
5.	B	15.	D	25.	A	35.	B
6.	D	16.	C	26.	D	36.	C
7.	C	17.	D	27.	D	37.	B
8.	B	18.	A	28.	C	38.	D
9.	A	19.	A	29.	D	39.	D
10.	D	20.	B	30.	D	40.	A

SECTION B

1. lithium   metallic (or metal)

   boron   covalent   network or lattice

   nitrogen   (discrete) molecular (or molecule) or diatomic

2. (a) (i) 8

   (ii)

   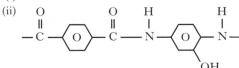

   (b) dissolves (or soluble) in water

3. (a) (i) rate of forward reaction equals rate of reverse reaction or concentration of reactants and products remain constant

   (ii) decreases (or reduces or gets smaller or diminishes or lowers)

   (b) no. of moles $= \dfrac{0 \cdot 010}{32} = 3 \cdot 125 \times 10^{-4}$

4. (a) they react with the oxygen (or are oxidised) or burn or react to form CO_2 or CO

   (b) $Q = I t = 50\,000 \times 20 \times 60 = 6 \times 10^7\ C$

   Al $3 \times 96\,500\ C \longleftrightarrow 1$ mol

   $6 \times 10^7\ C \longleftrightarrow \dfrac{6 \times 10^7 \times 27}{3 \times 96\,500} = 5596\ g$

5. (a) (i)

concentration of reactants	volume of reactants
(or permanganate or oxalic acid)	(or permanganate or oxalic acid)

   (ii) colour change is too slow (or too gradual or takes a long time) or colour change is indistinct

   (b)

   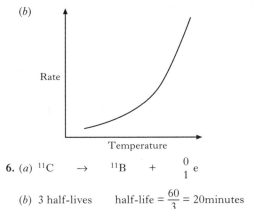

6. (a) $^{11}C \rightarrow {}^{11}B + {}^{\ 0}_{\ 1}e$

   (b) 3 half-lives   half-life $= \dfrac{60}{3} = 20$ minutes

   (c) ^{11}C

   more ^{11}C atoms or more radioactive atoms or greater mass of ^{11}C or ^{11}C has no other elements

7. (a) intermolecular attractions (or forces) or attractions between molecules

   (b)

   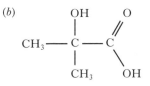

8. (a) (i) a reactant from which other chemicals can be made (or synthesised or produced or obtained or derived) or product of one reaction becomes the reactant of another

   (ii) addition (or additional)

   (iii) sodium chloride

   (iv) fats and oils are renewable (or will not run out or are unlimited) or propene is obtained from a finite source

   (b) $2C_3H_8O_3 \rightarrow 3CO_2 + 3CH_4 + 2H_2$

   (c) $3C + 3O_2 \rightarrow 3CO_2 \quad -394 \times 3 = -1182\ kJ$

   $4H_2 + 2O_2 \rightarrow 4H_2O \quad -286 \times 4 = -1144\ kJ$

   $3CO_2 + 4H_2O \rightarrow C_3H_8O_3 + 7/2\ O_2 = +1654\ kJ$

   addition $= -672\ kJ\ mol^{-1}$

9. (a) carbon, oxygen, nitrogen and hydrogen

   (b) count the number of (oxygen or gas) bubbles produced in a given time or measure the volume of gas produced in a given time or measure height of bubbles (or foam) produced in a given time or find rate of gas production

   (c) increasing temperature can denature the enzyme or idea of optimum temperature

10. (a) for drying, entry delivery tubes must be below surface of concentrated sulphuric acid and exit tube must be above

   for collection, apparatus must be workable and 'cooler' labelled, eg use of an ice/water bath

   (b) 1 mol $SO_2 \rightarrow$ 1 mol SO_3

   64·1g $\rightarrow$ 80·1g

   51·2 tonnes $\rightarrow \dfrac{51 \cdot 2 \times 80 \cdot 1}{64 \cdot 1} = 64 \cdot 0$ tonnes

   % yield $= \dfrac{\text{actual}}{\text{theoretical}} \times 100 = \dfrac{43 \cdot 2}{64 \cdot 0} \times 100 = 67.5\%$

   or

   moles of $SO_2 = \dfrac{51 \cdot 2}{64 \cdot 1} = 0 \cdot 799$

   moles of $SO_3 = \dfrac{43 \cdot 2}{80 \cdot 1} = 0 \cdot 539$

   % yield $= \dfrac{\text{actual}}{\text{theoretical}} \times 100 = \dfrac{0 \cdot 539}{0 \cdot 799} \times 100 = 67.5\%$

11. (a) (i) outer electron is further away from the nucleus or greater number of electron shells

   and

   (increased) shielding (or screening) by the inner electrons or decreased nuclear attraction due to inner election shells

   (ii) $3 \cdot 94 \times 10^{-21} \times 6 \times 10^{23} = 2371 \cdot 9\ kJmol^{-1}$

   (b) $Cl(g) + e^- \rightarrow Cl^-(g)$

12. (a) moles of $LiOH = 0 \cdot 1 \times 0 \cdot 4 = 0 \cdot 04$

   moles of $CO_2 = \dfrac{0 \cdot 24}{34} = 0 \cdot 01$

   0.02 mol of $LiOH$ reacts with 0.01 mol of CO_2

   excess $LiOH = 0.02$

   (b) 13

   (c) two points related to weak acid equilibrium two points related to water equilibrium

   or

   salt of a weak acid and a strong base for one mark

13. (a) (i)

$$CH_3 - \overset{\overset{\displaystyle CH_3}{|}}{\underset{\underset{\displaystyle CH_3}{|}}{C}} - \overset{\overset{\displaystyle CH_3}{|}}{\underset{\underset{\displaystyle H}{|}}{C}} - CH_3$$

    (ii) all have branched-chains (or branches)

(b) (i) more complete combustion (or less incomplete combustion or less CO) or higher octane rating or burns more smoothly or prevents knocking (or auto-ignition), or carbon burns more cleanly or reduces the oxygen (or air) required for combustion

    (ii) any correct ether isomer

(c) cyclohexane or any correct cyclic isomer

14. (a) (i) 2. measure the temperature (of the water)
        4. measure the highest temperature reached by the solution

    (ii) to reduce (or prevent) heat loss to the surroundings or to keep heat in or less energy lost (or to conserve energy)

    (iii) 1 mol KOH = 56·1 g

        $1\cdot2\,g \longleftrightarrow 1\cdot08\,kJ$

        $56\cdot1 \longleftrightarrow \dfrac{1\cdot08 \times 56\cdot1}{1\cdot2} = -50\cdot49\,kJ\,mol^{-1}$

(b) enthalpy change is for the formation of **one** mole of water or equivalent

15. (a) **x** is O-H   **y** is C-H

(b) (i) condensation or esterification
    (ii) 2 peaks only: at 1705-1800 and 2800-3000

16. (a) *Any 2 from:*
    flask should be swirled
    read burette at eye level
    white tile under flask
    add drop-wise (near end-point)
    no air bubble in burette
    use an indicator to give a sharp colour change
    rinse with solutions being used
    titrate slowly
    remove funnel from burette
    put a piece of white paper behind burette
    stir constantly, etc.

(b) (i) no. of moles of MnO_4^- (aq) $= 21\cdot6 \times 1\cdot50 \times 10^{-5}$
                                $= 3\cdot24 \times 10^{-4}$

    mole ratio 2:5

    no. of moles of $NO_2^- = 8\cdot1 \times 10^{-4}$

    concentration $= \dfrac{8\cdot1 \times 10^{-4}}{0\cdot025} = 3\cdot24 \times 10^{-2}\,mol^{-1}$

    (ii) $NO_2^-(aq) + H_2O(l) \rightarrow NO_3^-(aq) + 2H^+(aq) + 2e^-$

CHEMISTRY HIGHER 2011

SECTION A

1.	D	11.	C	21.	B	31.	B
2.	A	12.	C	22.	A	32.	D
3.	A	13.	A	23.	C	33.	D
4.	D	14.	D	24.	B	34.	A
5.	B	15.	C	25.	D	35.	B
6.	D	16.	D	26.	D	36.	C
7.	B	17.	D	27.	B	37.	C
8.	C	18.	A	28.	B	38.	A
9.	C	19.	C	29.	D	39.	B
10.	A	20.	B	30.	C	40.	A

SECTION B

1. (a) Homogeneous

(b) (i) Answer 0.0015
    (ii) New line should start **at same point as original** and should have a **steeper gradient**

2. (a) (i) more protons or increasing nuclear charge
    (ii) $Cl(g) \rightarrow Cl^+(g) + e^-$

(b) Argon does not form (covalent) bonds
    or
    No electrons involved in bonding

3. (a) Covalent bonds not being broken
    or
    Intermolecular bonds that are breaking

(b) Formula refers to the ratio of $Mg^{2+}:Cl^-$ ions (in lattice) (or alternative wording ie in the lattice there are twice as many chloride ions as magnesium ions)
    or
    Mg^{2+} ions surrounded by > 2 Cl^- ions
    or
    Cl^- surrounded by >1 Mg^{2+}

4. (a) 2,2,4-trimethylpentane

(b) It has more volatile (compounds)/vaporise more easily
    or
    (hydrocarbons) boil more easily/lower boiling point
    or
    more short chain compounds/lower GFM/more butane
    or
    Less viscous

(c) *Any four from the following for a maximum of two marks*
    ½ mark for safe heating method (no flame)/water bath
    ½ mark for condenser of some type
    ½ mark for methanol and stearic acid or "reactants"
    ½ mark for (concentrated) sulphuric acid in test tube
    ½ mark for pouring the mixture into a carbonate solution or solid carbonate added after esterification

5. (a) 1 mole $Ca(OCl)_2 \rightarrow 2$ moles Cl_2

    143g $\rightarrow$ 48 litres

    $\dfrac{0.096}{48} \times 143$

    $= \underline{\textbf{0.286g or 0.29g}}$

    or

    moles of Cl_2 $\dfrac{0.096}{24} = 0.004$

    moles of $Ca(OCl)_2$ $\dfrac{0.004}{2} = 0.002$

mass of Ca(OCl)2 = 0.002 × <u>143g</u>

              = <u>**0.286g**</u>

(b)

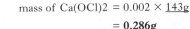

6. (a) $CH_3-CH_2-CH-CH_3$
                   |
                  OH

(b) ½ mark for triethanol amine has <u>hydrogen bonds</u> (between the molecules)
½ mark for triisopropyl amine molecules has van der Waals/or permanent dipole/permanent dipole attractions or doesn't have H-bonds
½ mark for H-bonds strong(er) (than the dipole/dipole)
½ mark for more energy/higher temp required (to overcome/break intermolecular forces)

7. (a) $C_8H_9NO_2$

(b) amino acids

(c) 0.0225 or 0.022 or 0.023
(can be rounded to 0.02 if working given)

8. (a)

<!-- polymer structure -->
$$\left[\!-C(=O)-C_{10}H_6-C(=O)-O-CH_2-CH_2-O-\!\right]_n$$

or

$$-C(=O)-C_{10}H_6-C(=O)-O-CH_2-CH_2-O-$$

(b) **EITHER**
1 mole glycerol → 1 mole ethane-1,2-diol
    92g   → 62g
27.6kg → 18.6kg

% yield = $\dfrac{13.4}{18.6}$ × 100

% yield = 72 %

or

moles of glycerol = $\dfrac{27600}{92}$

moles of glycerol = 300

actual moles ethane-1,2-diol = $\dfrac{13400}{62}$

actual moles of ethane-1,2-diol = 216.13

% yield = $\dfrac{216.13}{300}$ × 100

% yield = 72 %

9. (a) Palm oil has lower degree of unsaturated/palm oil less unsaturated/palm oil more saturated/palm oil contains more saturates/fewer double bounds
    or
Molecules in palm oil can pack more closely together

(b) Polyunsaturated

(c) Soap/emulsifying agent/detergent/washing/cleaning

10. (a) (i) $O_3 + 2KI + H_2O \rightarrow I_2 + O_2 + 2KOH$
    (ii) purple or blue/black or black or blue

(b) ½ mark for power supply/battery/lab pack
½ mark for (dilute sulphuric) acid labelled
½ mark for method for collecting O_3 which would work at positive electrode

(c) ½ mark for O_3 collected
(i) acidified dichromate (solution)
(ii)

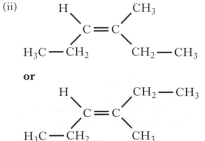

or

11. (a) Partially ionised/not completely dissociated

(b) (i) Contains more H^+ ions/higher concentration of H^+ ions
(ii) Because it is diprotic/dibasic/has two hydrogens
Or balanced equations for the reactions
Or sulphurous acid has more hydrogens
Or sulphurous acid has a high power of hydrogen
(iii) pH = 13

12. (a) Neutron to proton ratio (is unstable)
or proton to neutron ratio (is unstable)
or they have too many/few neutrons

(b) $^{131}_{53}I \rightarrow \,^{131}_{54}Xe + \,^{0}_{-1}e$

$^{131}_{53}I \rightarrow \,^{131}_{54}Xe + \,^{0}_{-1}e^-$

$^{131}I \rightarrow \,^{131}Xe + e^-$

$^{131}I \rightarrow \,^{131}Xe + e$

$^{131}I \rightarrow \,^{131}Xe + \beta$

(c) (i) 8 days
(ii) ½ mark for correct data from graph 70
½ mark for conversion to mole (÷ 131) 5.343×10^{-13}
½ mark for use of 6.02×10^{23}
½ mark for answer 3.22×10^{11} (ions)

13. (a) On addition of NaOH(s)…
- OH^- react with H^+
- concentration of H^+ decreases
- equilibrium position to shift to the left
- CrO_4^{2-} ion concentration increases

Solution becomes more yellow/less orange

(b) (i) • Mention of washings/rinsings
• Make the (standard) flask up to the mark with water/add water until desired volume reached
(ii) **EITHER**
moles $FeSO_4$ $0.02 \times 0.0274 = 0.000548$

moles of CrO_4^{2-} $\dfrac{0.000548}{3} = 0.000183$

Concentration of CrO_4^{2-} $\dfrac{0.000183}{0.050}$

                     = 0.00365 or 0.004 (mol l^{-1})

or

Candidates may use a "titration" formula of which an example is shown below.

$$\frac{C_1 V_1}{b_1} = \frac{C_2 V_2}{b_2}$$

$$\frac{C1 \times 50.0}{1} = \frac{0.0200 \times 27.4}{3}$$

$$C_1 = \frac{0.0200 \times 27.4}{3 \times 50.0}$$

Concentration of CrO_4^{2-} = 0.00365 (mol I^{-1})

14. (a) Answer within range -2640 to -2690

(b) $E = mc\Delta T = 0.2 \times 4.18 \times 40 = 33.44$

74g gives $33.44 \times 74 = 2475/2477$ kJ

Enthalpy of comb. $= -2475/-2477$

(c) Reversing given equation $+354$
using enthalpy of combustion of $C-5 \times 394$ or -1970
using enthalpy of combustion of $H_2 -6 \times 286$ or -1716

Addition -3332

15. (a) precipitation

(b) compound Z is water, **or** H_2O **or** steam **or** hydrogen oxide

(c) (i) The chlorine gas produced during the electrolysis of cerium chloride can be recycled/ reused (back into stage 4)

or

a substance may be added to reduce the temperature at which $CeCl_3$ melts

or

$CeCl_3$ can be electrolysed in solution (to avoid heating costs for $CeCl_3(l)$ electrolysis)

(ii) $Q = It$

$Q = 4000 \times 10 \times 60$

$Q = 2400000C$

$Ce^{3+} + 3e^- \rightarrow Ce$

$3 \times 96500C \rightarrow 140.1g$

$2400000 C \rightarrow 1161.45g$ or 1.16 kg

16. (a) 3

(b) $0.204°C$

CHEMISTRY HIGHER 2012

SECTION A

1. D	**11.** D	**21.** A	**31.** B
2. C	**12.** B	**22.** B	**32.** D
3. D	**13.** B	**23.** B	**33.** A
4. C	**14.** C	**24.** A	**34.** C
5. B	**15.** D	**25.** C	**35.** B
6. D	**16.** B	**26.** D	**36.** D
7. C	**17.** A	**27.** D	**37.** D
8. B	**18.** A	**28.** C	**38.** A
9. A	**19.** C	**29.** B	**39.** B
10. C	**20.** A	**30.** A	**40.** C

SECTION B

1. (a) Boron or Carbon or B or C or graphite or diamond

(b) Number of protons <u>increases</u>
or <u>increased</u> atomic number
or <u>greater</u> nuclear/positive charge
or <u>greater</u> pull on (outer) electrons

2. (a) To prevent loss of any solution/spray/acid from flask
or
To allow gas to escape
or
Spurting
or
To stop any <u>solids/liquids</u> getting in/out

(b) (i) 0.017
(ii) Answer between 0.37 and 0.4

3. (a) $E_h = cm\Delta T$
$= 4.18 \times 0.5 \times 82$
$= \pm171$ kJ

Number of moles required $= \dfrac{171}{1367}$

Answer 0.12 or 0.125 or 0.13 moles

(b) *Any two from:*
heat lost to surroundings
incomplete combustion (of alcohol)
ethanol impure
loss (of ethanol) through evaporation

4. (a) $^{89}Sr \longrightarrow {}^{89}Y + \beta$
or
$^{89}_{38}Sr \longrightarrow {}^{89}_{39}Y + {}^{0}_{-1}e$

(b) (i) No effect/no change

(ii) $\dfrac{89}{160} \times 10 = 5.56g$ or 5.6g

(c) ¼ **or** 0.25 **or** 25%

5. (a) (24 litres) $24,000$ cm$^3 \rightarrow 6·02 \times 10^{23}$

(0·110 litres) 110 cm$^3 \rightarrow 110/24000 \times 6·02 \times 10^{23}$
$= 2.76 \times 10^{21}$

(b) $CH_3CH_2OH + O_2 \rightarrow CH_3COOH + H_2O$

or

$CH_3CH_2OH + O_2 + 4H^+ + H_2O \rightarrow 2H_2O + CH_3COOH + 4H^+$

(c) Catalyst/reactants different state

6. (a)

or any structure for an expansion of the shortened structural formula $CH_3S_2CH_3$ containing
- 6 hydrogen atoms, valency 1
- 2 carbon atoms, valency 4
- 2 sulphur atoms, valency 2 or 4 or 6

(b) (i) *Either:*

moles $\quad Cl_2 \quad 0.010 \times 0.0294 = 2.94 \times 10^{-4}$

moles $\quad H_2S \quad 2.94 \times 10^{-4}/4 = 7.35 \times 10^{-5}$

$conc^n \quad H_2S \quad \dfrac{7.35 \times 10^{-5}}{0.05}$

$\qquad\qquad = \underline{1.47 \times 10^{-3}}$

or

Candidates may use a "titration" formula of which an example is shown below.

$\dfrac{c_1 v_1}{b_1} = \dfrac{c_2 v_2}{b_2}$

(ii) *This question is divided into two separate marks, the first subdivided:*

<u>First Mark</u>
Permanent dipole-permanent dipole attractions or polar-polar attractions/forces ½

weak intermolecular bonds/forces ½

<u>Second Mark</u>
If pd-pd named then:
Mention of difference in electronegativities or indication of polar bonds or indication of permanent dipole

If VdW/LDF named:
Instantaneous dipoles or temporary dipoles or uneven distribution of electrons or electron wobbles

7. (a) w=9 x=6 y=2 z=2
or
$C_9H_6N_2O_2$

(b) No elimination of a small molecule (such as water)

or The monomers have <u>added across</u> the (N=C) double bond

or Only one product molecule formed

or Joined across the (N=C) double bond

(c) Dotted lines between H/N or H/O on adjacent polymer chains

8. (a) Amide link or peptide link or peptide bond

(b)

(c) Essential

(d) 1 mark for wet paper towel (condenser) or cold finger test tube
1 mark for use a condenser
1 mark for raise the test-tube so that a greater length of the test-tube is above the hot water, but with the reaction mix still immersed or lower the level of the water

9. (a) ½ mark for bromine (water)/iodine (solution)
½ mark for Oleic decolourises
or stearic does not decolourise/decolourises slowly

(b) Octadec –9, 12, 15 –trienoic acid
Octadeca –9, 12, 15 –trienoic acid

(c) **EITHER** O^-Na^+ **or** CO^-Na^+ **or** COO^-Na^+ **or** O^- **or** $C-O^-$ **or** COO^-

10. (a) Air

(b) Methyl methanoate

(c) **EITHER**
$HCOOH \quad \rightarrow \quad HCONH_2$
1 mole $\qquad\qquad$ 1 mole
46g $\qquad\qquad\quad$ 45g
1.38g $\qquad\qquad$ 1.35g

% yield $= \dfrac{0.945g}{1.35g} \times 100$

$\qquad\qquad = 70\%$

or
moles HCOOH $\rightarrow 1.38/46 = 0.03$
moles $HCONH_2 \rightarrow 0.945/45 = 0.021$

$HCOOH \quad \rightarrow \quad HCONH_2$
0.03 moles $\quad \rightarrow \quad$ 0.03 moles

% yield $\qquad = 0.021/0.03 \times 100$

$\qquad\qquad = 70\%$

11. (a) (i) 3-methyl butan-2-ol

(ii)

(b) (i) $4BF_3 + 3\,NaBH_4 \rightarrow 2B_2H_6 + 3\,NaBF_4$

(ii) $-36\,kJ$
$-1274\,kJ$
$3 \times -286 = -858\,kJ$

½ mark for each correct enthalpy change

½ mark for addition of 3 sensible numbers

$-2168\,kJ\,mol^{-1}$

(c) 143444 **or** −143444 **or** 143000 **or** −143000 **or** 145000 **or** −145000
or
143 MJ

12. (a) To allow the potato discs/catalase to reach the pH of the buffer
or
To allow buffer to soak/diffuse into the potato disc
or
To allow the enzyme/potato to reach the same pH as the surrounding solution
or
To allow the enzyme/potato to acclimatise

(b) hydrogen peroxide/H_2O_2

(c) The enzyme is denatured
or
The enzyme changes its shape
or
Enzymes work best at an optimum pH
or
Too acidic for enzyme to function

13. (a) $Q = I \times t = 5.0 \times 60 \times 32 = 9600$ C

1 mol F_2 needs 2 moles of electrons = $\underline{2 \times 96\,500\,\text{C}}$

193000 C → 38g
9600 C → 1.89g

(b) (i) exothermic or heat given out or ΔH is −ve or $\Delta H < 0$

(ii) Graph shows as pressure increases/conc^n C_2F_4 decreases.

Line sloping <u>downward</u>

(c) depletion/break down of the ozone layer

14. (a) (i) $[H^+(aq)] = 1 \times 10^{-5}$ mol l^{-1}

(ii) *The marks for this question are divided into two separate marks:*

The first mark is awarded for the ammonia/ammonium equilibrium:

½ mark for
$NH_3\,(aq) + H_2O\,(\ell) \rightleftharpoons NH_4^+\,(aq) + OH^-(aq)$

1 mark if it has been shown that the position of this equilibrium is such that the ammonium ions tend to remove OH^- ions from solution
e.g. $NH_4^+\,(aq)\,OH^-\,(aq) \rightarrow NH_3 + H_2O$
or suitable description in words

The second mark is awarded for the water equilibrium:
½ mark for
$H_2O\,(\ell) \rightleftharpoons H^+\,(aq) + OH^-\,(aq)$

1 mark if it has been shown that water molecules dissociate resulting in an increased H^+ ion concentration
e.g. $H_2O\,(\ell) \rightarrow H^+\,(aq) + OH^-\,(aq)$
or suitable description in words

(b) Answers showing an appreciation that a <u>large volume</u> or large number of moles of gas is produced
or
There is an increase in the number of moles of gas
or
Oxygen gas is produced which can support combustion
or
It is an oxidising agent

15. (a) (i) to keep the current constant or to adjust the current

(ii) the current and the time

(b) (i) Recycle/reuse the <u>SO_2</u> and/or <u>H_2O</u>
or O_2 can be sold

(ii) $H_2O \rightarrow H_2 + \frac{1}{2}O_2$ **or** $2H_2O \rightarrow 2H_2 + O_2$

16. (a)

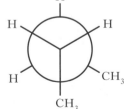

(b) 2-methylbutane
or
methylbutane

Hey! I've done it

BrightRED
PUBLISHING

Published by Bright Red Publishing Ltd, 6 Stafford Street, Edinburgh, EH3 7AU
Tel: 0131 220 5804, Fax: 0131 220 6710, enquiries: sales@brightredpublishing.co.uk,
www.brightredpublishing.co.uk

Official SQA answers to 978-1-84948-284-4
2008-2012